Wading for Bugs

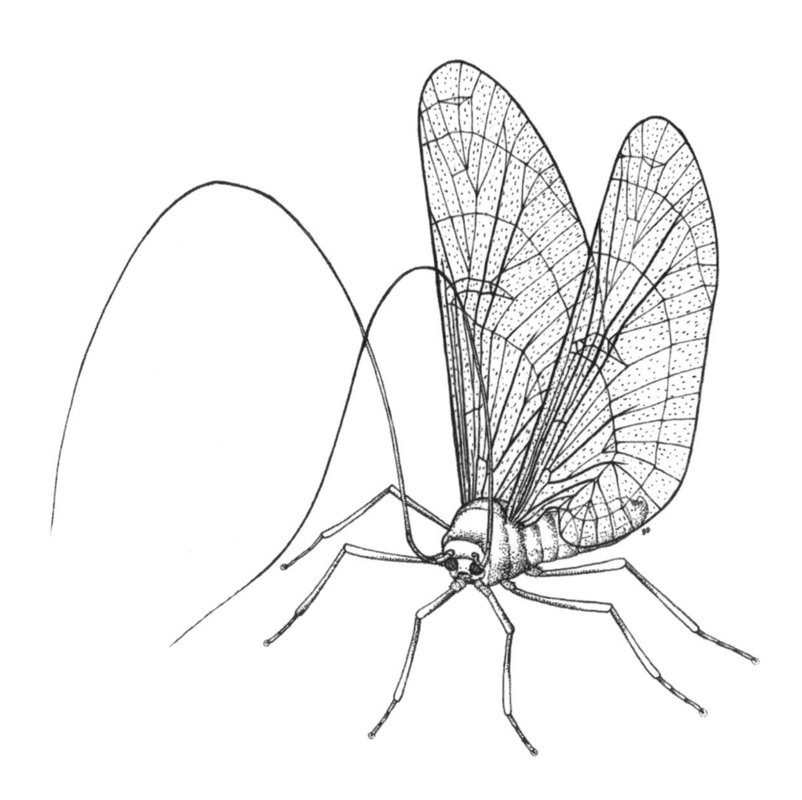

Wading for Bugs

Exploring Streams with the Experts

Edited by Judith L. Li and Michael T. Barbour

Illustrated by Boonsatien Boonsoong

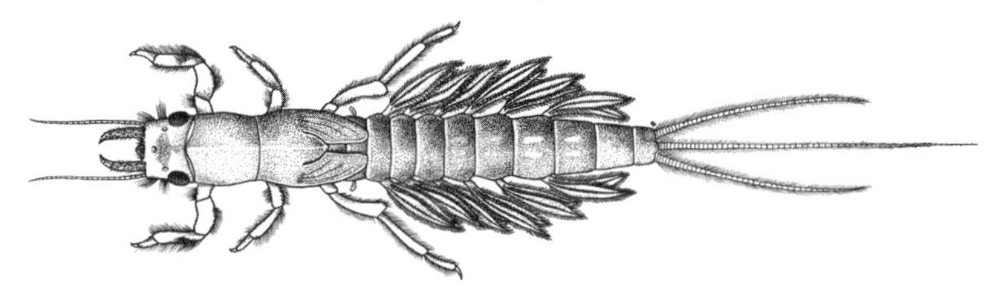

Oregon State University Press • Corvallis

The paper in this book meets the guidelines for permanence and durability of the Committee on Production Guidelines for Book Longevity of the Council on Library Resources and the minimum requirements of the American National Standard for Permanence of Paper for Printed Library Materials Z39.48-1984.

Library of Congress Cataloging-in-Publication Data
Wading for bugs : exploring streams with the experts / edited by Judith L. Li, Michael T. Barbour.
p. cm.
Includes bibliographical references and index.
ISBN 978-0-87071-608-9 (alk. paper) -- ISBN 978-0-87071-643-0 (ebook)
1. Aquatic insects. 2. Aquatic biologists--Anecdotes. 3. Aquatic habitats. I. Li, Judy. II. Barbour, Michael T.
QL472.W33 2011
595.76--dc23

2011029079

First published in 2011 by Oregon State University Press
Printed in the United States of America

Oregon State University Press
121 The Valley Library
Corvallis OR 97331-4501
541-737-3166 • fax 541-737-3170
http://oregonstate.edu/dept/press

In appreciation of their commitment to and stewardship of our vital waterways, we dedicate this collection of stories to the legions of volunteers and professionals who monitor and work to restore streams and rivers in their communities.

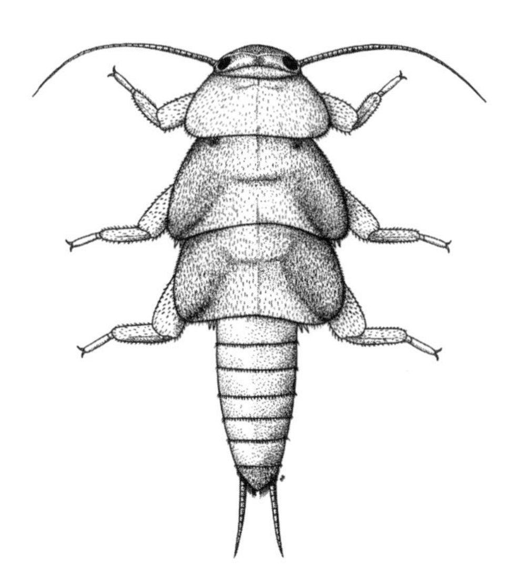

Contents

Introduction xii
Judith Li and Michael Barbour

I Mayflies in the Mist

About Mayflies (Ephemeroptera) 1

1 The Ghost Mayfly 4
John Woodling

2 Emergence of a Mayfly: Nuisance or Blessing? 11
Michael T. Barbour

3 Mayflies and Fly Fishing at the Forks of the Credit 15
Lynda D. Corkum

II Stonefly Recoveries

About Stoneflies (Plecoptera) 22

4 A Cosmic Stonefly: Rediscovering *Tallaperla* 25
Dave Penrose

5 Returning Salmonflies to the Logan River 29
Mark Vinson

III Sleuthing for Caddis

About Caddisflies (Trichoptera) 34

6 A Case that Glitters 37
Vincent Resh

7 Life in a Cornucopia 42
John Richardson

8 Mystery of the Spine-Adorned Caddisfly 48
Marilyn Myers

9 Dicos for Ducks 52
Judith Li

10 Digging in a Ditch for Caddis 57
Norman H. Anderson

11 A Criminal "Case" Made with Caddisflies 61
John R. Wallace and Richard W. Merritt

IV Truly Flies

About True Flies (Diptera) 65

12 Encounter with Arctic Black Flies 68
Donna Giberson

13 Hanging on in the Alpine Tundra 71
Deb Finn

14 Making the Case for an Aquatic Insect and Its Habitat 79
Richard W. Merritt

15 Way Cool Mountain Midges 83
Gregory W. Courtney

16 Marine Sea Stars, Nudibranchs, and Midges 89
David Wartinbee

17 The Phantom Midges of Silver Lake 94
Michael C. Swift

V Dragonfly Detectives

About Dragonflies (Odonata) 100

18 Hanging from a Leaf 103
Rob Cannings

19 Tracking the River Cruiser 108
Christopher Beatty

VI Bugs and Beetles on Their Best Behavior

About Beetles (Coleoptera) 112

About True Bugs (Heteroptera) 114

20 In Defense of Whirligig Beetles 117
Fred Benfield

21 The Bugs Famously Known as Ferocious Water Bugs, Giant Water Bugs, and Toe Biters 123
David A. Lytle

22 Riding the Current for the Riverine Backswimmer 127
Michael Bogan

23 Secrets of an Infrequent Flyer 132
Mark P. Miller

Anatomy of a Mayfly 137

Biological Assessment:
Using Biological Indicators to Evaluate the Health of a Waterbody 138

Glossary 142

Useful References 145

About the Contributors 147

Index 155

Illustrations

Ephoron adult 4

Ephoron nymph 10

Hexagenia imago 11

Hexagenia nymph 13

Baetis tricaudatus 15

Ephemerella subvaria 15

Paraleptophlebia adoptiva 20

Tallaperla 25

Pteronarcys nymph 29

Pteronarcys adult 32

Neophylax rickeri larva in case 37

Neureclipsis larva 42

Neureclipsis case in leaf 46

Dicosmoecus larva in case 52

Dicosmoecus larva 52

Pseudosteophylax larva in case 57

Pycnopsyche larva in case 61

Simulium adult 68

Simulium larva 73

Metacnephia larva 74

Bibiocephala larva 79

Bibiocephala adult 82

Deuterophlebia adult 83

Deuterophlebia larva 88

Tanytarsus (Chironomidae) larva 89

Chaoborus larva 94

Stylurus adult 103

Stylurus nymph 107

Macromia magnifica adult 108

Macromia magnifica nymph 111

Dineutus adult 117

Dineutus adult, side view 118

Abedus 123

Martarega 127

Ambrysus 132

Baetis nymph, outline 137

Stylurus stalking fly larva 144

Acknowledgments

We are grateful for the advice provided by two anonymous reviewers of earlier drafts, and we are particularly appreciative of guidance by Mary Braun and Jo Alexander at the OSU Press and editor Joanna Conrad in the production of our book. Thanks also to Jeremy Monroe at *Freshwater Illustrated* for his efforts in finding the perfect photograph for the cover, and to Charlie MacPherson, who facilitated the delivery of our book's illustrations from Thailand. The editors and most of the contributors are members of the Society for Freshwater Science (formerly the North American Benthological Society), and we have greatly appreciated the network of colleagues that organization provided as we developed this collection of stories.

Introduction

Whether turning over stream rocks, chasing damselflies at stream's edge, or watching the dance of mating mayflies, children and adults alike have long delighted in the diversity of stream insects. If you also like dabbling in creeks or poking around in streams, undoubtedly you've plucked a rock out of the water and discovered a miniature community of insects scrambling around underneath. Sharing a similar fascination for these organisms, aquatic biologists work to figure out where those critters live, how they survive, and eventually how they are distributed within our water systems. Aquatic biologists can tell a lot about the health of a stream by seeing what lives there. This book is a collection of stories relating the adventures, misfortunes, and surprises that we aquatic entomologists have encountered as we uncover fascinating details about stream insects. Scattered across the North American continent like the widely dispersed organisms they study, the contributing authors tell us about insects living in exotic and commonplace habitats. Some stories were revealed slowly over time, while others were discovered quite abruptly.

The energy and enthusiasm generated in these stories reveals our amazement at how stream insects are adapted for the conditions where they live, how they behave, and what they require to survive from one generation to the next. Often the secrets to survival lie in the unique characteristics of each taxon. For example, some insects have anatomical specializations that are specific to particular conditions: gills along the abdomen can help with respiration, forelegs can act like filtering fans, or sucker-like appendages can keep an insect attached to rocks. In addition, flexible behaviors like burrowing into the streambed, drifting with the flow, or emerging out of the stream also demonstrate adaptations to life under changing conditions. To help you compare what you see in your environment with the array of insects described in our stories, we share our experiences with the hope that you and other lovers of streams find in our stories a joy reflecting your own experiences.

Each chapter of this book focuses on a particular aquatic insect or a group of closely related ones. The chapters are arranged according to the taxonomic order to which those particular insects belong. This organization is in keeping with the traditional inclinations of biologists, but more importantly, it provides an opportunity for you to appreciate similarities and recognize differences between related organisms. To help you get acquainted with the insect orders, we begin each section with a description of what the order's members generally look like, typical life histories, and use as bioindicators. Illustrations included in each chapter will help you recognize similarities and differences within each order. The stories told by the authors illustrate personal, often humorous, ways in which they came to appreciate the idiosyncrasies of particular insects. A few are told in a fictional context, but the biological details are accurate. Within any insect order, some taxa may be found everywhere, while others might be quite specialized. For example, consider the common October caddisfly, *Dicosmoecus,* compared to the unusual caddisfly *Pedomoecus*, which specializes in the sands of desert springs. Some taxa fit our stereotypical associations (such as the fiercely biting *Simulium tormentor* black flies in the Arctic), while others are quite unique (as is the rare blackfly *Metacnephia coloradensis* in the high Rockies).

If you have already been introduced to tools of bioassessment, you may be familiar with numerical values that come from designations of invertebrates as indicators (for national standards established by the U.S. Environmental Protection Agency in their Rapid Bioassessment Protocols (RBPs), see p. 138). Professionals who conduct monitoring and assessment programs use biological indicators to determine the ecological condition of waters running through their jurisdiction; many watershed interest groups also use these same tools. The expression "EPTs," representing the mayflies (Ephemeroptera), stoneflies (Plecoptera), and caddisflies (Trichoptera), may be part of your new vocabulary, or perhaps the abbreviation doesn't conjure up any biological significance. Our chapters about each of these insect orders will help you understand why many of these insects are regarded as fairly sensitive stream inhabitants.

The array of mayflies at the Credit River, the cosmic stonefly, and the net-building *Neureclipsis bimaculata* are good examples of such sensitive species. We hope the stories in our volume—along with information about each taxon's juvenile and adult stages, feeding practices, and preferred habitat conditions—will help you interpret the meaning of biological indicators.

Despite all our efforts to generalize broadly over many environments and geographic regions, the incredible variety of insects defies absolute, definitive classification. Among our stories are several indicator-defying examples like the non-biting black fly in "Hanging on in the Alpine Tundra," or the riverine, sediment-loving mayfly described in "Emergence of a Mayfly." We hope you will begin to develop your own bioassessment skills as you use biological indicators in your local stream or watershed.

How to Use This Book

The stories collected here are meant to entertain and to educate. The book is arranged by the taxonomic orders where the insects belong; general information about each order precedes the stories, and specific information about characteristics of the individual insects discussed follows each story. We have tried to simplify terminology, and we have also provided an appendix about bioindicators and a glossary, both of which are intended to help further your understanding of insect biology and aquatic ecology.

We hope you will enjoy this collection of stories as much as we do.

Judith L. Li and Michael T. Barbour

PART I

Mayflies in the Mist

About Mayflies (Ephemeroptera)

As the Latin name for this order of insects implies, mayflies are ephemeral as adults, when they dance in the air above streams seeking their mates. They spend most of their lives as nymphs molting through many stages, until they emerge, only to reproduce, lay eggs, and die within days. Timing is everything for such brief mating encounters, and emergence of adult mayflies is often synchronized within a population to ensure that males will find reproductively mature females. The males emerge first and begin their swarming dance; later, females enter the swarm of males, and they mate in flight. All this occurs without feeding because adult mayflies have only remnant, non-functional mouthparts and do not feed.

Morphology

Immature mayflies are called larvae, or nymphs. Their head shapes vary from ones shaped much like those of grasshoppers, somewhat ovoid in cross-section, to very flattened forms that may be coin-like in appearance. Nymphs have moderately large eyes and well-developed antennae, and their heads can have a variety of bumps, projections, or armature. The thorax has three distinct segments, each bearing a pair of legs; wing pads that grow as the nymph matures are attached on the second and third thoracic segments. The most distinctive features of mayfly larvae are their abdominal gills; they acquire vital oxygen through gill surfaces. Unlike the gills of stoneflies, those of mayflies occur almost entirely on the abdomen. Mayfly gill morphology varies greatly; gills can be feathery, plate-like, dissected, or fringed in single or layered combinations. Most mayflies have three "tail-like" filaments, called caudal filaments, extending from the last abdominal segment. In some species the middle filament is quite reduced or absent, and it should not be relied upon as a dependable trait for distinguishing between mayflies and stoneflies (stoneflies have two filaments).

The subimago stage of a mayfly occurs just when the larva emerges out of the water. During this brief interval (only a few hours), reproductive organs and wings mature while the subimago warms up, dries out, and remains perched quietly upon the water's surface or on streamside plants or rocks. This is a very vulnerable time because the mayfly's immobility makes it susceptible to invertebrate or vertebrate predators. Compared to the fully formed adult, the subimago's wings are poorly defined, venation is incomplete, and coloration is generally dull. A dense layer of fine hairs, called microtrichia, cover the subimago's body and wings, creating a very dull appearance. Microtrichia are waterproof, and that thick layer protects the nymph as it emerges out of the water.

The mayfly molts one last time to the adult (imago) stage, during which it is ready to fly and mate. In comparison to its subimago stage, the adult has fully veined wings and brighter colors. In most species the adult bears two pairs of wings: the larger forewings and much smaller hind wings. To perform their graceful mating dances, adults use their somewhat triangular wings, held perpendicular to the thorax, while long, caudal filaments are upturned and held high over their backs. Adult males usually have very large, often bright-orange, eyes that help them spot their mates.

Life History

Many mayflies have only one generation per year (termed univoltine), but multivoltine species (that have several generations per year) are also common. All mayflies undergo gradual changes after hatching from eggs and as they grow from the nymph to subimago stage. Diapause (when growth is suspended) likely occurs during the egg phase for most North American mayflies. Mayflies molt as many as thirty times, shedding their external skeletons at each molt to accommodate their expanding size. There are not distinctively different morphologies in the immature nymphal phase (such as a pupal stage), but later instars have wingpads that grow with each molt until the nymph emerges. The number of molts, which varies with each species, can also change with environmental conditions such as temperature or food resources. Optimal conditions often result in the minimal number of molts for the species. Many larval mayflies move downstream by intentionally lifting off the stream bottom and drifting with the current. Others may be washed into the flow when stream discharge is high. After the brief subimago stage, the adult lifespan lasts from a few hours to three days. Females often lay eggs by quickly dipping their abdomens into the stream. Individuals are sometimes caught in the current and swept into the drift as they swim to the surface to emerge or as they return to the water to lay eggs.

Bioindicators

Mayflies (Ephemeroptera), partnered with stoneflies (Plecoptera) and caddisflies (Trichoptera), are part of the triad of excellent bioindicators known as EPTs. When mayflies are diverse in an aquatic system, that is, when there are many species present, the stream is considered to be in good condition. Most mayfly species are sensitive to human disturbances such as chemical contamination or habitat alteration. When certain species that aquatic biologists expect to be in a particular place are missing (oftentimes determined by a reference condition), their absence is an indication that the ecosystem is impaired. However, some mayflies are less sensitive than others; for example, some burrowing mayflies are quite tolerant of silty river bottoms. Thus interpretation can be misleading if more tolerant species are counted as if they were evidence of more healthy conditions.

References

Edmunds, G. F. Jr., S. L. Jensen, and L. Berner. 1976. *The Mayflies of North and Central America.* Minneapolis: U. of Minnesota Press.

Waltz, R. D., and S. K. Burian. 2008. "Ephemeroptera." In *An Introduction to the Aquatic Insects of North America*, edited by R. W. Merritt, K. W. Cummins, and M. B. Berg, 181–85. Dubuque, IA: Kendall Hunt.

1 The Ghost Mayfly

John Woodling

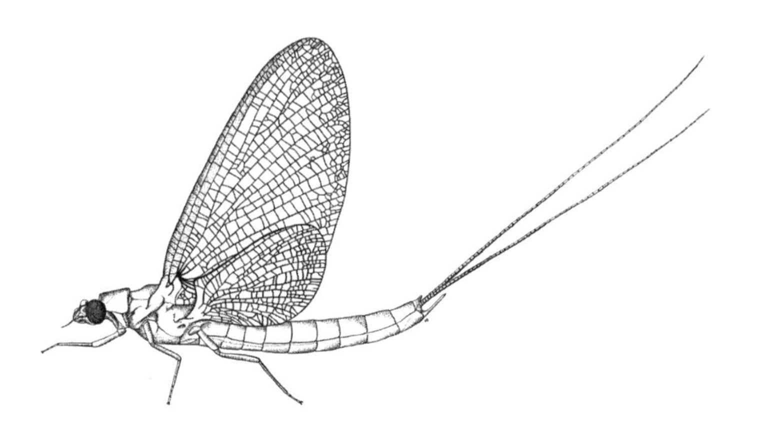

Anglers and stream biologists know that flowing water and ever-changing rivers create experiences both rare and beautiful. Sometimes they present questions that may never be answered. I remember one foggy evening on July 27, 1969, when Jerry Parsons and I, both graduate students at the University of Louisville, were collecting adult aquatic insects from Brashears Creek in Spencer County, Kentucky. A slight change in barometric pressure created a light fog in the Brashears Creek Valley, bothering what few drivers were out and about. Enough light remained so that we noticed a fog bank rapidly moving towards us, winding through the overhanging trees and shrubs and curling over the stream. The sun had set, and dusk was deepening. To lure flying adults, we'd suspended a white collecting sheet from a tree limb to reflect light from our ultraviolet blacklight lamp. We were hoping to match adults with larval and nymph stages from the stream, but that night the few caddisflies and stoneflies we attracted were the same as others we'd seen for the last few weeks.

Without a breath of wind, the fog rolled over and past us. We watched streamers of mist weave through the limbs and leaves of the pin oaks lining both sides of the creek. Instantaneously the temperature dropped several degrees, although we didn't feel cooler. It was Kentucky: the evening remained hot and muggy. The flying insects also didn't seem bothered by the temperature change. Bugs continued to fly into the sheet, creating a light patter of sound as wings, bodies, and legs hit the cloth. We talked idly and drank from our half-gallon bottle of Pisanno, a cheap red wine that accompanied us on those lazy evening collecting jaunts. The fog quieted the few country sounds, though a soft gurgle of water flowing over riffle cobblestones wafted in the swirling mist.

Gradually we became aware of faint plopping sounds coming from the stream. Jerry played the beam of his flashlight onto the water. We saw fish rising and feeding on pale white insects that floated on top of the slowly swirling stream. These white wisps were about an inch and a half long with wings only slightly shorter. The white mayflies were emerging to mate, floating down the stream using their cast nymphal husks as rafts while their wings inflated and dried. They were a movable feast for the fish. With a sweep of our butterfly net, we got a few of the first mayflies rising up into the air. They were very delicate and had useless legs that couldn't support their bodies. To our surprise, this was a species completely unknown to us.

Quickly, the numbers of flying mayflies increased, and the sound of fish feeding intensified. Evidently the change in pressure and temperature had triggered a mass emergence of the pale mayflies. Jerry and I completely lost interest in using the blacklight, which didn't matter since the mayflies weren't attracted to the light. The mass of flying pale insects formed an undulating tube that floated silently over the stream through the fog, rising and falling from a height of about ten to fifteen feet down to the surface of the stream. Like a ghostly vapor, the entire swarm stayed over the water, never straying past the adjoining stream banks. The pale, silent cloud of life wove through the evening mists, confined by the darkening tree mass along the stream bank. We stood mesmerized by the spectacle.

A moment later, many of the mayflies looked like two individuals flying in tandem. As the lead portion of an individual rose and fell, the trailing part followed like the rear end of a roller coaster. We used our trusty butterfly nets to grab a few. What looked like a pair of individuals hooked together was actually a single mayfly. Because most mayflies are subimagoes, not true adults, when they first emerge from the water, they must molt one more time before mating. What we saw as the front part of the tandem pair was an adult, or imago; the back part was the cast skin of the subimago, which remained attached to the two tails of the adult. These mayflies had only a few seconds to wiggle out of their subimago skins before rising out of the water, assisted only by useless legs. Neither Jerry nor I could find any white mayflies molting anywhere on the stream bank or in the riparian vegetation, where sometimes we'd seen other species going through this change to their last life stage. Somehow these white mayflies cast their skin while flying through the air.

When we heard a very faint clicking sound in the sky above the swarm, Jerry's flashlight pinpointed a large brown insect flying erratically through the sky: its beating wings were creating the clicking sound. Looking much like a World War I fighter plane, the insect tipped over and flew right down the flashlight beam, landing on Jerry's arm. I gingerly picked the insect off his arm and carefully placed it in the killing jar. The creature was a female Dobson fly, *Corydalus cornutus.* We could tell its gender because the pincers on the mandibles were smaller than a male's would have been. That evening this female, before it was enveloped by the mayfly swarm, had been looking for a suitable mate of its own. (All these years later, I still have the Dobson fly in a display case over my desk. I look at it as I type these words.) While the fish continued to feed on emerging mayflies, we pulled a minnow seine from the back of our Jeep. One seine haul showed that smallmouth bass, rock bass, sunfish, and minnows, evidently representative of the fish in the stream, were all gorging on the mayflies. Their stomachs were full and rock hard, yet the fish kept eating, with their mouths full of mayfly adults and nymphs. A sunfish the size of my hand had a stomach the size and shape of a golf ball. Still the fish continued to feed until mayflies got caught in their gill arches and were coming out the back of the gill flaps. Then, in less than an hour, the swarm diminished. We watched several mayfly pairs mate and lay eggs. The fish were finally sated. The surface of the stream was once again smooth and silent, except for the corpses of spent mayflies, swirling on the currents. After the wine was gone we loaded up and left. Only the fog, the stream, and the dead mayflies remained.

The next day we identified the white mayfly as *Ephoron leukon* (Williamson) using a classic text published in 1953: "The Mayflies or Ephemeroptera of Illinois," by B. D. Burks. This identification created consternation in the lab. We were creating a species list of the aquatic community in the Salt River drainage (of which Brashears Creek was a tributary); to that end, we had collected 177 benthic aquatic insect samples from Brashears Creek, and about two to three times that many from the Salt River. Not one *E. leukon* nymph was present in all those samples—none, nada, zip. Now we had a new mission: to find and capture the elusive and mysterious *E. leukon* nymphs.

There was no great mystery about this mayfly's biology. In the 1930s, F. P. Ide from the University of Toronto discovered that only the *E. leukon* male molted in flight. The female did not molt prior to mating, an observation supported by the preserved mayflies we had collected. Ide also described nymphs of this species inhabiting small tubes constructed of silt in riffles. Our sampling device had collected insects only in the upper two to three inches of the stream substrate: the white mayfly obviously lived much deeper than where we had been sampling. The next spring, prepared with shovels and nets, we dug down into the cobble until we hit limestone bedrock under the riffles. Still, we didn't find any of those lovely white mayflies. We sampled the bottom of pools covered in silt—no nymphs. We even dug into the stream banks at and below the water line—no nymphs. We caused more than a little change in the stream's substrate, sampling for several months, but never collected a single nymph. The white mayfly was a ghost.

Ephoron leukon was obviously important in the stream. Given that fish metabolism increases as temperature increases, and using temperature data from Brashears Creek, we figured that the fish obtained about 1.5 percent of their total annual energy from the one-hour hatch of *E. leukon* on July 27th. According to

D. B. Burks, this species emerged several times over the course of any summer. If it emerged from three to four times a year, the "ghost" mayfly could supply more than 5 to 6 percent of the total available energy for fish at Brashears Creek. But our knowledge of the species' distribution was as shallow as the riffles from which we took samples.

After graduation, I moved from Kentucky. I never found nymphal *E. leukon* in Brashears Creek. From that point forward, every time I thought I knew something about aquatic insects, the memory of the fog and the white ghosts rolled back into my mind.

Thirty-four years later, I was driving across the Colorado River in Grand Junction, Colorado, more than 1,300 miles from Brashear's Creek, a retired fish biologist no longer responsible for stream studies. Looking out over the river like I always did while crossing a bridge, my eyes picked out a swarm of white floating above the water. Many thousands of white insects gently floated over the river. I stopped in the middle of the road, got out of my pickup, and walked over to the side of the bridge. The September evening was beautiful; the temperature had dropped several degrees in a rather short time, perhaps triggering the emergence of both these white insects and my memories. I did not have a bug net, but I did have a large plastic cup filled with an iced drink—diet Coke this time, not Pisanno. I dumped out the drink, got back behind the wheel, and drove slowly back across the bridge, holding the cup outside over the roof of the pickup cab. Sure enough, a few white mayflies got caught. Once again I was looking at white mayflies about one and a half inches long with very weak legs.

The next day, the insects and I went to a friend at the Colorado Division of Wildlife. I borrowed a dissecting microscope and identified the white mayflies as *Ephoron album* (Say) with the help of *The Mayflies of North and Central America* (a key by Edmunds, Jensen, and Berner that was written in 1976). I learned that the genus *Ephoron* only has two species. I had now collected both species as adults, but neither as nymphs.

The following Wednesday I was scheduled to give a presentation to Dr. Russ Walker's techniques course in stream biology at Mesa State College. We were going to a place along the Colorado River about five miles upstream from where I captured the adult mayflies. Perhaps if my student "volunteers" sampled diligently, with a lot of effort we might collect a few *Ephoron* nymphs. At our sampling reach, the Colorado was mostly a long run with fist or larger size cobbles on the streambed and lots of silty sediment filling the spaces between the rocks. We collected dozens of mayflies in every seine haul. Evidently *E. album* built nests in the silt and did quite well.

For the next two years, during September, I frequently saw late-afternoon swarms of the white mayfly over the river and even over large irrigation canals throughout the Grand Valley. We observed in the techniques class each fall that, far from being ghosts, these mayflies seemed to be everywhere. Thus, in 2008 I confidently told Russ's class that we could expect to collect lots of ghost mayflies, a species that was hard to catch anywhere except near Grand Junction. We collected none. Not a single individual white mayfly was collected that year. *Ephoron* was a ghost once again.

Why did the white mayfly in both Kentucky and Colorado seem to vanish? I remembered a vicious downpour that hit Kentucky in March of 1970, the spring after we'd spotted the ghost mayflies. Ten inches of rain fell in the area of Brashears Creek, transforming it from a stream that could be easily waded to more than twenty-five feet deep for a day or so. Likewise in the spring of 2008 before our futile collection on the Colorado, snowmelt in western Colorado was extensive and very high. Remember that the ghost mayflies burrow in silt: high flows would have removed the silt where *Ephoron* grew as larvae, possibly accounting for the precipitous drop in their numbers. Return of the ghost mayflies likely came with recovery of their silty habitat.

To solve the mysterious comings and goings of the ghost mayflies, we had to be at the right place at the right time. Truly the study of streams can be an art form as much as a science.

We should pursue and demand creative reflection, tenacity, and a long-term commitment to place instead of relying on an illusory technique based only upon a few samples. Establishing creative and cooperative endeavors among stream biologists and volunteers may best serve all of us and the rivers we need and love.

Ephoron leukon and Ephoron album

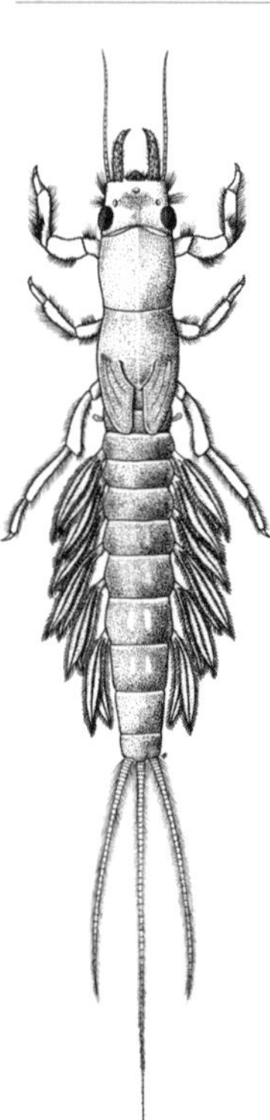

Life History: Univoltine.
Larvae: Nymphs build tubes in thick silt.
Adults: White subimagos emerge in autumn. Males molt to adult stage in flight; females do not have final molt until mating.
Feeding: Larvae collect small particles that float through tubes they build in fine silt. Adults do not feed.
Habitat Indicators: These mayfly species reflect the thick, stable silt conditions required to maintain the tubes built by their nymphs. They are found in large rivers, and their emergence is very sensitive to changes in temperature. *Ephoron* nymphs build a case of sand and silt among rocks. When silt conditions are suitable, they occur in high densities and can be a significant food resource for fish.

2 Emergence of a Mayfly: Nuisance or Blessing?

Michael T. Barbour

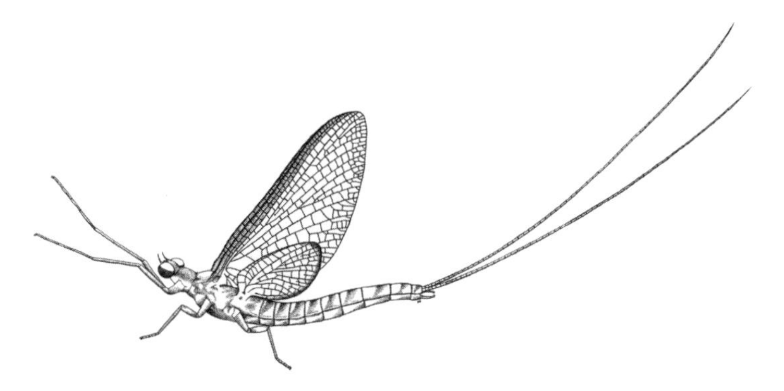

It was a full moon, and the summer night was hot and sticky. I listened to the night noises as we drove along the gravel road, the warm breeze coming in from the open window. My dad was humming a Gene Autry tune that was playing on the radio. The static sound coming from the dash didn't seem to bother Dad. He knew the words and filled in where the song was undecipherable otherwise. The old truck bounced along on the rough surface, jostling us constantly.

My thoughts turned to my upcoming birthday. In just six days, on July 21, 1954, I would be seven years old. I couldn't wait. When I turned seven, I would be as old as my best friend, Jake, who had turned seven two months ago. I hated being one of the youngest in my class, but I couldn't change that. I wondered what presents I would get. I was hoping I would get that Red Ranger BB gun I had seen in old man George's hardware store. But I knew that my Dad was not likely to be able to pay the high price of ten dollars to get me that gun. Maybe he would make me a slingshot instead. I had seen him putting an old bicycle tire tube in the shed just last week.

Dad brought me out of my revelry, exclaiming "Oh, my."

I glanced over and saw him staring ahead with a frown on his face. I looked out the dirty windshield and couldn't see anything—I mean *anything!* The moon was gone, the forest was gone, and the light from the truck was dimmed to almost no light at all.

All of a sudden, the windshield was being pelted with some unknown thing, or many things. Splat! Splat!

"Quick, roll up the window!" yelled my Dad.

I cranked the window knob as fast as I could, but many bugs flew in before I could get it closed.

"Dad, what are these bugs?" I cried. "Will they bite?"

"No, they are harmless. But they sure are pests."

"Why are there so many of them?"

"I don't know, son. This happens sometimes, and always in the summer. I've heard they come from the river. They come out all at once and fly around, getting in everyone's way. What a mess."

Dad stopped the car, then reached over and slapped at the bugs flying around inside the cab. I watched him swat at them until they were all dead.

We sat there for a long time, watching the dense cloud of winged insects attacking the truck. My eyes were wide with fear. I had never seen so many insects. I thought back to a bible story from ancient times when a swarm of locusts descended upon Egypt. These weren't locusts, were they?

Eventually, the swarming ended, and we heard no more bugs hitting the windshield and sides of the truck. Dad opened the door and stepped out. I hesitated only a minute before following him. As I stepped onto the running board, I heard a crunch and felt a gritty sensation through my sneakers. "Oooh," I exclaimed.

Dad ignored me and was standing in the front of the truck, where I joined him. He and I just stared at the grille of the truck, which was covered with insects. In fact, the whole front of the truck was outlined in insects. Some were crawling in a haphazard manner, but most were dead, or appeared so.

"I think these bugs are called shadflies or mayflies." Dad brushed the insects from the headlights, releasing the light that had been dimmed by the swarming. "They are good fish food. Your Uncle Frank and I will try to go fishing tomorrow. We should get a lot from the river—the fish will be biting."

I looked around at the layers of insects on the truck, on the ground, all over the road. All I could think of was to wonder where they all came from, and why there were so many of them.

Dad finally got the windshield cleaned up, and we went on our way. We both were silent after that. The moon was back. The heat of the summer night was almost unbearable. However, I kept the window up all the way home.

This is the story of my youth that my daughter enjoyed hearing over and over.

"Did you ever see bugs like those again?" she would ask.

"I never saw them in such great numbers as that summer," I would reply. "They emerged mostly in July, and there were many reports of *Hexagenia* emergence and swarming occurring in the 1940s and early 1950s around large rivers in the Midwest, particularly in water courses of Illinois and Wisconsin. But there weren't too many after that."

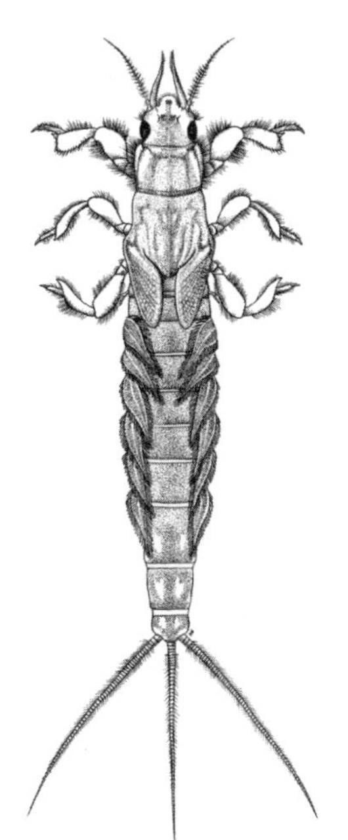

My daughter gave me a quizzical look.

I explained, "It wasn't until much later that I learned these swarms were a good thing. The insects are mayflies in the order Ephemeroptera, meaning 'lives but a day.' They spend most of their lives as juveniles in the water, and their emergence to adulthood is timed to occur within a short duration. Because the winged adults are short-lived, simultaneous emergence of males and females is important for producing eggs for the next cycle."

"But why is it good that these insects all come out at the same time and make a mess all over?"

"Because they indicate good-quality water. The mayfly juveniles, or nymphs, live in the bottoms of rivers and streams; if the water is healthy, they will be there in large numbers, burrowing into the mud. Those nymphs are an important source of food for fish. The ones that survive through many molts emerge to mate as adults and lay eggs."

"What has happened to them?" she asked.

"I'm afraid we haven't been treating our rivers as well as we should. We depend on those waterways for too many things, and we have polluted them in the process."

"Can we do something to help the rivers and the mayflies?"

"Yes, and we are already doing it. Our rivers are recovering in places where we have been restoring the habitat and treating water before it enters the waterway. In fact, we had a report that a mayfly emergence was detected recently in Wisconsin in such high numbers that the swarm showed up on a radar screen."

"That's good, isn't it?"

"Yes, that is very good."

Hexagenia spp.

Life Cycle: Univoltine; egg, nymph, subimago, imago are 4 stages of cycle.

Larvae: Known as nymphs; nearly 30 molts can occur before nymphs emerge after 1 year in the water.

Pupation: No pupation; nymphs emerge from water as subimagos; final molt to imago stage in a few hours.

Adults: Large imagos are dark red-brown, females lighter in color; at least 1 inch long; emerge in summer, usually at dusk; live for approximately 1 day.

Feeding: Nymphs are collectors of detritus and food particles; no feeding as adults.

Habitat Indicators: *Hexagenia* nymphs are burrowers in soft sediment. They have long, feathery gills to constantly move oxygenated water through their burrows. They are sensitive to pollution, particularly chemical and industrial waste; they require relatively clean sediments and low turbidity. Shifting sediments from habitat alteration is not conducive to maintenance of nymph burrows. These mayflies are important fish food, and their presence in large numbers is indicative of sustaining fish populations.

3 Mayflies and Fly Fishing at the Forks of the Credit

Lynda D. Corkum

To catch fish, anglers use artificial lures that imitate insects. Most aquatic insects, stoneflies (Plecoptera), caddisflies (Trichoptera), and midges (Diptera) included, have lures fashioned after them, but imitations of mayflies (Ephemeroptera) are by far the most popular lures used to catch trout. Although each mayfly species has a scientific name, most fly fishers know these insects by their common names. *Baetis tricaudatus* is known as a blue-winged olive, *Ephemerella subvaria* as a Hendrickson, *Paraleptophlebia adoptiva* as a blue quill, and the hefty *Hexagenia atrocaudata* as a late hex. These feathery, often glittery, artificial flies, tied on a hook using feathers, fur, and other flotsam, mimic the shapes and

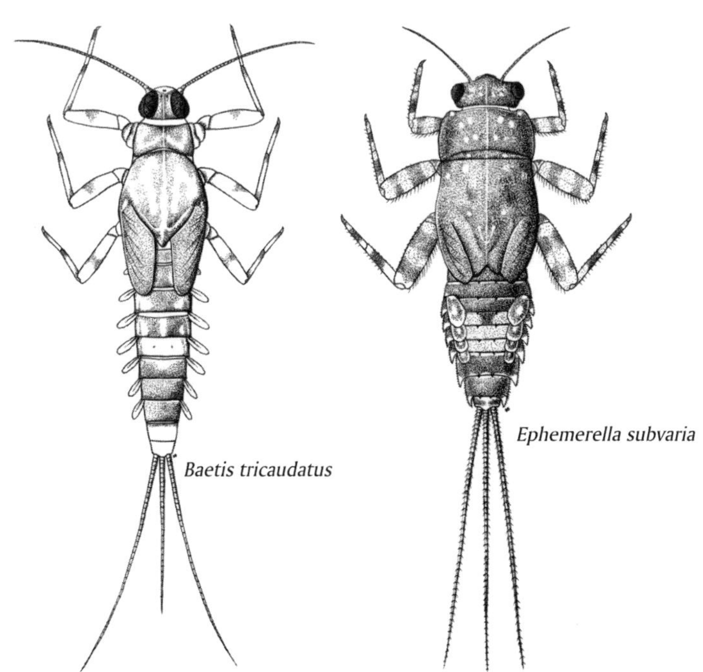

behaviors of mayflies moving through the water. Anglers entice trout to feed either on wet flies that sink, imitating nymphs, or on dry flies that float on the water's surface, copying adults just emerging or returning to lay eggs.

Live mayfly nymphs spend much of their lives crawling on the bottom of streams. When nymphs lift off or are swept off the stream bottom, they are whisked downstream by the flow; we call this "drifting" behavior. Trout tend to feed on drifting insects that are in the water or on the water's surface rather than on the bottom of the river.

In the mid-1970s, I went with Jan Ciborowski (now my husband) to the Forks of the Credit River in southern Ontario. There we found twenty-three species of mayflies—a wonderful assortment for studying drift. We chose this location because Dr. Fred Ide, a well-known aquatic entomologist, had recorded adult mayflies emerging "at the Forks of the Credit below the confluence and downstream near the bridge" many years before.

The Credit originates in the Niagara Escarpment, flows southward through mixed deciduous forests over limestone, and empties into Lake Ontario. The river cuts a valley into the escarpment, carving steep slopes into the rock; Jan and I followed a narrow, winding road along the river. The road took a hairpin turn near the village of Belfountain, where the west and east branches of the Credit, each about fifteen meters wide, come together, forming the Forks. Clear, fast-flowing water bubbled over rocky substrates and large woody debris—great habitat for trout and mayflies. The river was wadeable, but in a few spots was over fifty centimeters deep. These were terrific sampling conditions. After trying many sampling techniques, I discovered that simply picking nymphs off rocks and waterlogged branches with my forceps was the most efficient collecting method. I separated mayflies by species and placed them into thermos bottles filled with river water. They clung to the leaves I had added as a surface for them to hang onto for the trip home. All season long, Jan and I collected our daily ration of mayfly nymphs and transported them for laboratory studies at Erindale College (now known as the University of Toronto at Mississauga).

In the lab, I wanted to know what made different mayflies leave the streambed and enter the water column. It was easy to manipulate the number of nymphs, substrate type, and current flow in our artificial streams. With a simple adjustment of a motor-driven paddle wheel, I changed the speed of the water that circulated around the elliptical raceway. Hundreds of observations and many experiments later, we learned how drift habits were related to a mayfly's general behavioral type.

Baetis tricaudatus nymphs live in rapidly moving water and sit on fallen branches or on rock surfaces, especially moss-covered areas exposed to flowing water. These "swimmers" drift in the water and parachute to the stream bottom as the current slows. They are minnow-like and have a streamlined body and a round head. In our experiments, these mayflies preferred fast water. Many more swimmer nymphs entered the drift at low flows than at high flows.

Ephemerella subvaria nymphs cling to rock surfaces in medium- to fast-flowing water, but when dislodged they will "dog-paddle" back and forth in the water. They didn't drift as often as "swimmers," but they certainly drifted more than the "crawlers" in our lab streams. Typical of "clinging" mayflies, *Ephemerella* are squat and stocky. The gills are on the back of the abdomen in contrast to many other mayflies that have gills along the sides.

Paraleptophlebia adoptiva have a flattened body and head. Forked gills are along the sides of the body. We found them on the underside of rocks or fallen branches in slow to moderate flowing water. When they were in quiet waters, *Paraleptophlebia* nymphs crawled out from underneath to sit on the tops of rocks. They are dubbed "crawlers." As their names suggest, crawling mayflies prefer slow-moving water; they entered the drift more at higher than at lower velocities. Overall, far fewer crawling nymphs drifted from the substrate in our raceway than swimming or clinging species.

Hexagenia are "burrowing" mayflies with tusks that are modified mouthparts. These tusks help them dig in the mud-clay river bottom to form U-shaped burrows. Within the burrows, the body of a *Hexagenia* nymph undulates and the feathery gills beat slowly. No surprise that burrowing mayflies seldom entered the water column of a stream until emergence.

Each mayfly adult stage is short lived, hence the name Ephemeroptera (ephemera=temporary, living but a day; optera=wing); but there is variation in the time for eggs to hatch and in the time for nymphal growth (see Table 1). These variations in egg and nymphal development determine the period of time for adult emergence. For example, *Baetis tricaudatus* has two generations each year. The slow-growing, over-wintering nymphs emerge into adults in the spring and the fast-growing generation emerges in late summer. The two emergence periods of this species, along with the different times when other mayflies emerge, extend the time period when trout seek out adult mayflies to eat. The longer emergence periods also lengthen the window when anglers can hope to trick the fish into taking their carefully placed flies. Once, in late autumn, I had been completely focused on my sampling, but when I lifted up my head to stretch, I discovered the first snowfall of the season had begun. To truly capture the differences in emergence, I was collecting all year round.

The Credit is a popular stream for anglers, and I was careful never to bring a rod and reel with me. During my frequent trips, some guardians of the river would check to see that I wasn't rushing the fishing season. Calling down from the bridge, they'd ask, "Are you collecting rocks?" They saw me sitting on a river boulder, head down, picking up rocks, cobbles, and branches. I must have looked a bit odd.

Table 1. Emergence and Hatching Times for Four Mayflies.

Species name	*Behavioral type*	*Emergence**	*Hatch time*
Baetis tricaudatus	swimmer	April–May; August	afternoon
Ephemerella subvaria	clinger	April–May	afternoon
Paraleptophlebia adoptiva	crawler	June–July	late morning to midday
Hexagenia atrocaudata	burrower	July–August	after sunset

* at Forks of the Credit River

Taking the rock I was working on, I walked to the river's edge to explain. "No, not rocks, but look at the insects I'm collecting."

One spring afternoon, I met a Scottish fly fisher on the Credit. He studied the river and told me there was a brown trout at the east branch just upstream from where it joins the west branch. Off he went, walking quietly across the river, careful not to disturb his target. Within moments, he caught a large brown trout. He held it up for me to see and then gently released it back into the river. I could see the Hendrickson on the end of his tippet! Obviously, he had the perfect fly for that spring day (see Table 2).

Not long ago I received fly-tying supplies, feathers, hair, and other treasures that had belonged to our advisor, Dr. Phil Pointing, who had been an avid fly fisher. These days, Jan and I are still thrilled to wade in running waters, excited not only by what we've learned about stream mayflies, but also by the challenge of catching trout with the perfect fly.

Baetis tricaudatus; Ephemerella subvaria; Paraleptophlebia adoptiva

(See Chapter 2 for *Hexagenia* spp.)

Life Cycle: *B. tricaudatus*: bivoltine; *E. subvaria, P. adoptiva*: univoltine. Stages of cycle: egg, nymph, subimago, imago.

Larvae: Nymphs that shed their skins as they grow; can have as many as 30 molts, but often less, depending on food resources, temperature, and other growing conditions.

Pupation: Development is gradual, and there are no pupae. When fully grown, the nymphs swim or crawl to the water surface and emerge to the winged sub-adult (or subimago) stage to rest while reproductive organs mature. A subimago has cloudy, smoky wings and is called a dun.

Adults: After the final molt from sub-adult stage, sexually mature adults (imagos) emerge. The imago has clear, shiny wings and is called a spinner by fishermen. Sub-adults and adults have no mouthparts and do not feed. Each stage lasts about a day. After mating, females fly over the river to deposit eggs on the river. Females of some species dive under the water to deposit eggs. Males and females die shortly after mating.

Table 2. Dry Flies (Adult Imitations).

	Blue-wing olive
Species imitated	*Baetis tricaudatus*
Hook size	14–22
Body of dun	brownish-olive
Wing of dun	gray
Tail of dun	2
Body of spinner	brownish-olive
Wing of spinner	clear
Tail of spinner	2

	Hendrickson
Species imitated	*Ephemerella subvaria*
Hook size	12–14
Body of dun	brownish-olive and yellow
Wing of dun	medium gray
Tail of dun	3
Body of spinner	reddish-brown
Wing of spinner	clear
Tail of spinner	3

	Blue quill
Species imitated	*Paraleptophlebia adoptiva*
Hook size	18
Body of dun	reddish-brown
Wing of dun	slate
Tail of dun	3
Body of spinner	reddish-brown
Wing of spinner	clear
Tail of spinner	3

	Late hex
Species imitated	*Hexagenia atrocaudata*
Hook size	6
Body of dun	yellow, brown marks
Wing of dun	gray
Tail of dun	2
Body of spinner	yellow, brown marks
Wing of spinner	clear
Tail of spinner	2

Feeding: *B. tricaudatus* nymphs are collectors of algae and small food particles.

E. subvaria nymphs feed on decomposing plant material.

P. adoptiva nymphs feed on small bits of decomposing leafy materials or algae.

Habitat Indicators: As described in this story, mayflies occur in a variety of habitats, from swiftly flowing streams to slow, muddy rivers. Because they have exposed gills, most mayflies require moderate to high levels of oxygen and unpolluted waters. Their use as water quality indicators works best when they are classified to at least family levels.

B. tricaudatus nymphs are one of the most widespread mayfly species in North America. The nymphs are found in erosional areas of streams where water flows over gravel and rocky bottoms. They are also associated with plants and occur on wooden branches or logs that have fallen into rivers. These nymphs commonly drift at night.

E. subvaria nymphs dwell in rocky bottoms of riffles in mid-size to large rivers throughout eastern North America. The nymphs are more active (they wiggle) in slow water and tend to hold their position (cling to the substrate) in fast water. The nymphs frequently drift at night.

P. adoptiva nymphs live in gravel-bottomed shallow streams. The nymphs are often found under rocks in quiet water near shore. They crawl over substrates and are less prone to drift than *Baetis* or *Ephemerella*.

H. atrocaudata nymphs burrow in mud-clay sediments of rivers and beaver ponds. As expected, burrowing nymphs rarely occur in the drift until the substrate is disturbed. (See Chapter 2 for more information on *Hexagenia* spp.).

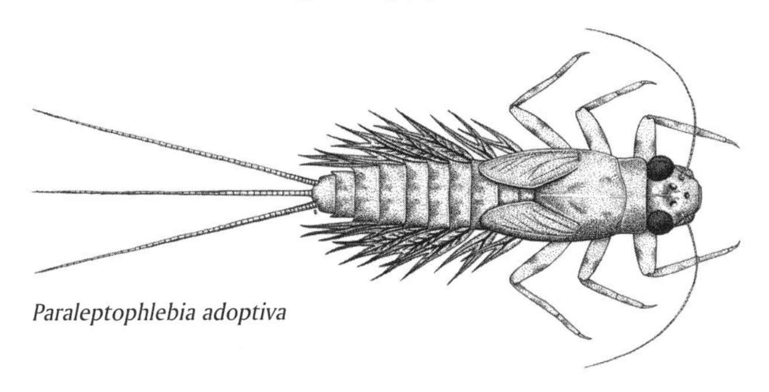

Paraleptophlebia adoptiva

PART II

Stonefly Recoveries

About Stoneflies (Plecoptera)

Compared to other more numerous and diverse insect orders, Plecoptera has relatively few members; nevertheless, stoneflies are important in aquatic systems. Because they generally live in well-oxygenated, cool streams (and sometimes lakes in similarly clear circumstances), they are very useful bioindicators. Plecoptera literally means "folded wings," describing how the hind wings tuck under the anterior wings when the adult stonefly is at rest. Their larvae (nymphs) are among the largest aquatic insects, and include the giant salmonflies (*Pteronarcys*) and golden stoneflies (family Perlidae) familiar to fishermen.

Morphology

Stoneflies have primitive morphological features that are relatively unspecialized. Body shapes include the stout, rounded peltoperlids with a roach-like appearance; thin, pencil-like winter stoneflies; flattened, marbled-colored golden stoneflies, and dark, flattened bodies of giant salmonflies that grow to two or three inches long. Herbivorous stoneflies often have larval mouthparts shaped for shredding leaves, and predacious species can stab and grasp their prey. The thorax has three distinct segments, each with a pair of crawling legs. Gills, either single or in tufts, often occur on the bottom (ventral) thoracic surface or along the bases of legs on the thorax; however some stoneflies are without gills, and a few have gills on the abdomen. Wings begin on nymphs as small pads on the second and third thoracic segments, growing longer with each molt. The abdomen has ten segments and a pair of somewhat stout tails (cerci) on the tenth segment.

Adult stoneflies carry their wings folded over their backs when at rest. They are not strong fliers and characteristically flutter in flight. Many species of stoneflies lack mouthparts as adults, and, like mayflies, those species do not feed.

Life History

Stoneflies commonly have one-year generation cycles (univoltine), though some live from one to two years (semivoltine), and occasionally two or three years. Stoneflies are flexible in the timing of their life cycles, varying rates of development through egg or nymphal life stages. Growth is gradual, in which nymphs metamorphose through a succession of molts (ten to twenty-two depending on species and conditions) until they are ready to emerge as adults. Stoneflies do not have a pupal stage. The stonefly's last molt is fairly unique among aquatic insects because the last instar nymph crawls out onto streamside rocks or vegetation for the final molt. Its skin splits along the top (dorsal) surface, and the soft-bodied adult basks in its new environment while its wings and body cuticle harden. Where stoneflies are abundant, gravels next to the stream may be covered in cast-off skins following the time when particular species emerge.

To attract mates, males of many species "drum" by tapping on a substrate (for example, tree bark or stones) with the tips of their abdomens. Virgin females respond with signals specific to each species. After mating, females release eggs on the stream surface, or deposit them into the water, where eggs attach to the substrates in the stream. In some species, particularly under unfavorable environmental conditions like drought, eggs undergo diapause (delayed development). In another strategy to avoid adversity, some stonefly nymphs burrow into gravels below the stream surface (a region called the hyporheos). Winter stoneflies (family Capniidae) often deploy this strategy, then emerging midwinter onto the snow and other winter conditions.

Bioindicators

Stoneflies are some of the best indicators of good water quality as they are among the most sensitive of aquatic insect larvae. Typically, they are among the first to disappear as pollutants enter an ecosystem. However, they are not useful indicators in warm, low-gradient streams because they are generally not present in those conditions.

References

Stark, B. P., S. W. Szczytko, and C. R. Nelson. 1998. *American Stoneflies: A Photographic Guide to the Plecoptera*. Columbus, OH: The Caddis Press.

Stewart, K. W., and B. P. Stark. 1988. *Nymphs of North American Stonefly Genera (Plecoptera)*. Lanham, MD: Entomological Society of America.

Stewart, K. W., and B. P. Stark. 2008. "Plecoptera." In *An Introduction to the Aquatic Insects of North America,* edited by R. W. Merritt, K. W. Cummins, and M. B. Berg, 311–13. Dubuque, IA: Kendall Hunt.

4 A Cosmic Stonefly: Rediscovering *Tallaperla*

Dave Penrose

Streams that drain the southern Appalachian Mountains in western North Carolina are an entomologist's playground. Most are swiftly flowing, clear, cold streams that support trout, including the native brook trout. Lots of stoneflies live there, too. During spring in the high-elevation streams of this region, the stout, roach-shaped *Tallaperla* is perhaps one of the most abundant stoneflies.

Leaves that fall into these streams are a primary source of energy for aquatic insects and fish. Like many other stream invertebrates in these headwaters, *Tallaperla* feeds by shredding leaf material. By consuming leaves, it brings terrestrial energy into the aquatic food web. When we want to find "shredders"—

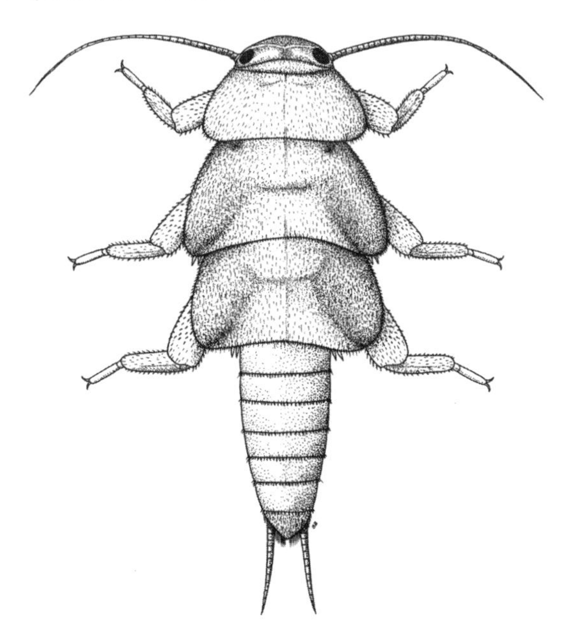

the aquatic invertebrates that eat leaves, woody twigs, or other organic materials—we search for leaf packs and places where these materials are retained. The leaves tend to "stack up" behind rocks and logs, making leaf packs where we often find *Tallaperla*. These habitats are plentiful in our southern Appalachian streams.

One spring, just as the leaves of the forest hardwoods were beginning to unfurl, the North Carolina Forest Service was alarmed to find gypsy moth larvae feeding on the tender new leaves. To control the outbreak, they deployed helicopters for broad-scale spraying of the insecticide Dimilin in affected watersheds. Dimilin adheres to the new leaves and acts as a gut crystalline when consumed, killing gypsy moth larvae but also other insects. After the forest service sprayed many stream basins across the state, it wanted to know if Dimilin was affecting non-target insects in streams. My colleague Mitchell and I were charged with answering their question.

In the streams we surveyed, most populations of *Tallaperla* had been eliminated after the Dimilin spraying. We thought the missing shredders in our survey made perfect sense: broad-scale spraying affected everything in the watershed. Dimilin was getting into the streams, adhering to the leaves in leaf packs, killing *Tallaperla* as well as the gypsy moths. Because of our study, the forest service changed their spraying strategy. They marked streams with helium balloons and told the helicopter pilots to shut off their delivery systems as they approached these balloons.

To monitor the recovery of the Dimilin-affected streams, Mitchell and I returned a year later and repeated our sampling. Interpreting complicated biological data to water quality managers, who are very rarely biologists, can be one of the greatest challenges for state regulatory biologists. For that reason, many states' biological monitoring programs use indices, metrics, and effective bioclassification systems to present data in non-technical ways. However, in North Carolina, many of the people we frequently met streamside not only understood these concepts, but also appreciated the diversity of life present in the streams we sampled. One such person was Judd.

Judd worked as a correctional officer and was not a state biologist, but he had a keen interest in stream life and wanted to help with the survey. Judd was impressed with what we picked up in our first sample.

"It's a stonefly, Judd, an insect that lives most of its life in streams around here," Mitchell said as they both studied the *Tallaperla* Judd held in his hand.

The forest service's strategy had worked! Indeed we found *Tallaperla* and other shredders in good numbers. Apparently there were enough of them in very small headwater reaches unaffected by spraying. These populations had reproduced and drifted into the stream reaches that had been poisoned the year before.

It was a good day to celebrate in our favorite bar, where Judd, who was still trying to figure out the stonefly story, joined us.

"That stonefly we found, is it like stickbait or the grampus we use for bait in the Swannanoa River?" Judd knew caddisflies (stickbait) and big hellgrammites (grampus) by names used by fellow fishermen.

"Well, not exactly," said Mitchell. "Stoneflies were better than stickbait."

Although Judd appreciated the recolonization of *Tallaperla* in the forest service stream, I watched him struggling with the concept. At that time I hadn't been a state stream biologist too long, but I quickly realized that this was a good opportunity. How could I help promote aquatic insects as tools to monitor water quality? How would I convince locals like Judd that insects living in the water can be good things? And then it hit me. There are almost two hundred species of stoneflies in streams in the southeast. But if *Tallaperla* was eliminated from this list of species, it would be like a canary in a coal mine—an indication that our entomologist's playground here in the southern Appalachians was less diverse by one species.

I thought about Mitchell's conclusion. Stoneflies were better? Not really, but Mitchell must have meant that stoneflies were more intolerant than most caddisflies or hellgrammites, making them good indicators of clean water. Better. I sat in the bar with the other two and sipped my beer.

We probably should have gotten some rest instead of barhopping, but that night we wanted to catch the Perseid meteor shower. Our little group bounced up the mountain in Judd's beat-up truck to an open spot on top, which happened to be a cemetery. I was pretty sure that Judd had been here before. It didn't take long to see our first meteor streaking across the open sky.

Judd pointed to it and shouted, "TALLAPERLA!!"

I had to smile. Cosmic stoneflies in this clear, cold Appalachian sky. Better.

Tallaperla spp.

Life Cycle: *Tallaperla maria*, common in the southern Appalachian mountains, has a semivoltine life cycle (1.5 years) and a 6–7 month egg diapause.

Larvae: Small, reddish-brown insects (7–10 mm) have a distinctive "roach-like" appearance. These insects are often the most abundant shredders in small mountain streams of the eastern United States.

Adults: Short-lived (1–4 weeks). Some species have small wings, but all are poor flyers; poor flight prevents them from crossing even small geographic barriers.

Feeding: Shredders of leaf material; diet includes detritus and diatoms contained in leaf packs.

Habitat Indicators: *Tallaperla* is extremely effective at and dependent upon shredding organic material. In many small Appalachian streams, *Tallaperla* may number 500 or more organisms per square meter of stream bottom. *Tallaperla* are good indicators of healthy habitats. Streams that become unstable cannot typically retain leaves and other organic material; they may build up fine sediments. Because food resources and desired habitat are lost, it is very likely that these disturbed streams will not support shedders in the benthos. However, if a stream is "restored" to a healthy system where streambed stability and retentive structures return, it should once again retain organic material; if it is a small Appalachian headwater, shredders should once again dominate the fauna. In many cases the presence of *Tallaperla* can be used as a criterion for restoration effectiveness.

5 Returning Salmonflies to the Logan River

Mark Vinson

Large, dark brown salmonfly nymphs make superb food for fish; those that survive to be winged adults may be eagerly snatched up by birds and bats in the riparian forest. Known to biologists as *Pteronarcys californica* (Plecoptera: Pteronarcyidae), they are also familiar to fishermen who watch for them to emerge in spring or summer. About twenty years ago I noticed that salmonflies were absent from the Logan River, a beautiful mountain trout stream located near Logan, Utah, where I lived. Why were these stoneflies found in nearly all neighboring streams but not in the Logan River? Ten years later, I learned, quite by surprise, that salmonflies were once the most common aquatic insect in that

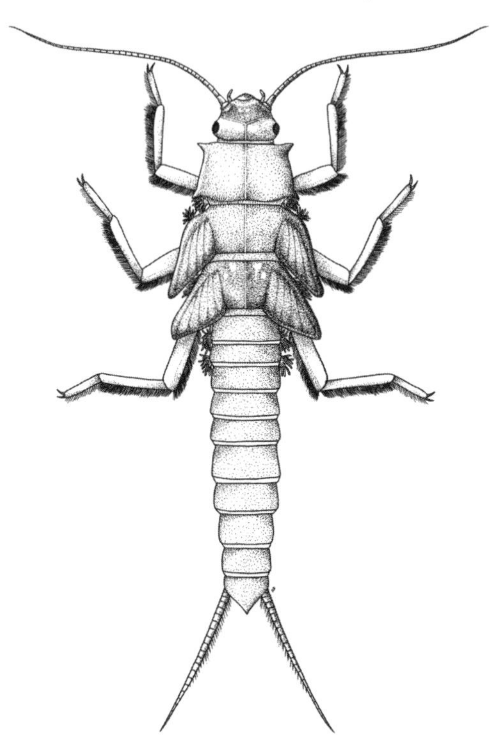

river. How could this be? Had they been there the entire time, though my students and I failed to collect them? This hardly seemed possible.

In 1927, James Needham, an eminent entomologist of his time, recounted collecting insects during his summer's stay in Utah the year before:

> *Pteronarcys californica* abounds in the clear waters of the Logan River below an elevation of 6000 feet. It is undoubtedly one of the most important insect species of the stream. Its greatest abundance seems to be in trash [debris] piles that gather against the upstream side of the larger rocks in midstream where it finds both food and shelter. Fifty or more well-grown nymphs could be taken on a screen by dislodging a single large stone. (Needham and Christenson, 1927)

Puzzled by the difference between my observations and those from years before, I spent the next decade searching for *Pteronarcys* along the edges and under stream rocks of the Logan River, especially in places near where Needham had reported their abundance. Dusty museum collections, long-forgotten theses, and agency reports confirmed that *Pternonarcys* had been abundant until the early- to mid-1960s. On September 7, 1966, Nancy Erman collected a few *P. californica* nymphs from the Logan River near the Mendon Bridge. Hers was the last known collection record from this river; after this, *P. californica* disappeared from the river's scientific record. Later investigators failed to collect this conspicuous stonefly, yet they made no mention of its absence. How could the most common, largest, and most charismatic aquatic insect in the river simply vanish from this spectacular habitat? Their disappearance appeared to go unnoticed, and the cause of their extinction seemed a mystery.

A highway chemical spill back in the early 1960s seemed probable, but no evidence for such an accident could be located in newspaper archives. Additionally, snow- and ice-melting chemicals and herbicides were applied along Utah Highway 89, which parallels much of the river. However, the Logan River's high-quality trout fishery and diverse aquatic invertebrate fauna (except the missing salmonflies) had not changed appreciably since the 1920s, so chronic pollution from those activities seemed unlikely.

My best, yet unproven, explanation for why salmonflies disappeared from the Logan River was broad-scale herbicide treatment for sagebrush eradication during the 1960s. Helicopters were used to spray the herbicide 2,4-D throughout the Logan River Basin between 1959 and 1969 to reduce sagebrush and improve grass forage for livestock. 2,4-D is highly toxic to animals that eat the sprayed vegetation. When leaves from streamside plants fall into the water, they provide a primary food source for many aquatic invertebrates, like salmonflies. Thus both terrestrial and aquatic invertebrates can be vulnerable to herbicide spraying. Given the time period when the spraying occurred, which coincided with when salmonflies disappeared from the record, the 2,4-D application appeared to be the cause of their original disappearance. But still baffling to me was why they hadn't returned over the last forty years. What else could prevent them from recolonizing? It was time to try some experiments.

For several years I placed salmonfly nymphs in cages throughout the Logan River watershed. *Pteronarcys* grow slowly and live three or four years as nymphs. Every few weeks I brought them decaying leaves for food and checked on their survival. They survived for nearly two years in the cages, so habitat conditions in the Logan River seemed quite suitable. Since they survived as nymphs, perhaps the difficulty in recolonizing was that salmonfly adults are poor fliers: dispersal by flying to new habitats may be slow. This hypothesis seemed to be corroborated by a study of salmonfly genetics in the Blacksmith Fork River, Logan River's largest tributary, which did indeed suggest *Pteronarcys* do not move around much (Schultheis et al. 2008).

Forty years after being eliminated from the Logan River, salmonflies still had not been able to reestablish on their own. But my lab and I decided to give them some help, and in 2004 we began reintroducing salmonflies. Each spring since then, just before salmonflies emerge, the Utah State University BugLab and dozens of Trout Unlimited volunteers collect thousands of mature salmonfly nymphs from the Blacksmith Fork River and release them into the Logan River. Thus far, the project seems to be successful. Now salmonfly nymphs can be found near our transplant locations. Every year more fishermen observe adults flying along the stream as the salmonflies look for a mate or a good place to lay their eggs. Luck and good stewardship in recent times were on our side. Because the river remains intact, *Pteronarcys* is slowly reestablishing in the Logan.

References

Erman, N. A. 1968. "Occurrence and Distribution of Invertebrates in Lower Logan River." Master's Thesis, Utah State University.

Needham, J. G., and R. O. Christenson. 1927. "Economic Insects in Some Streams of Northern Utah." *Utah Agricultural Experiment Station Bulletin* 201: 1–36.

Schultheis, A. S., J. Y. Booth, M. R. Vinson, and M. P. Miller. 2008. "Genetic Evidence for Cohort Splitting in the Merovoltine Stonefly

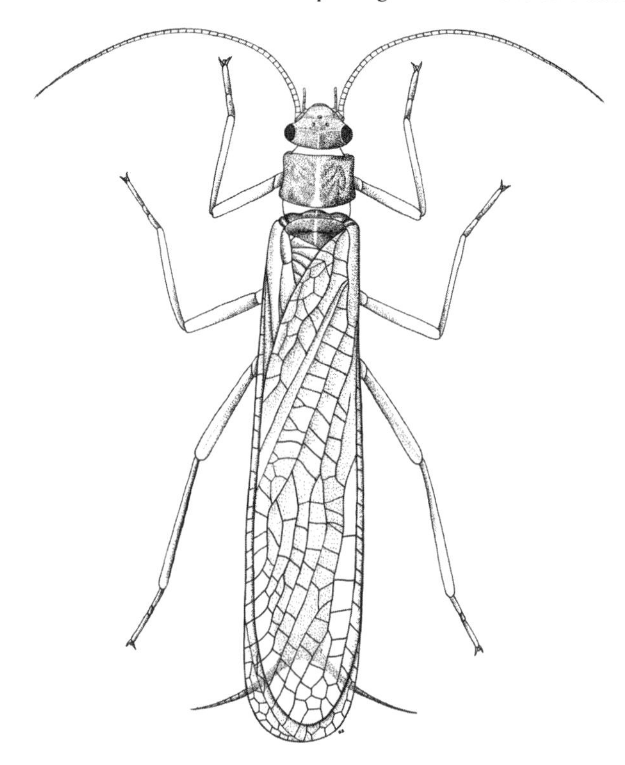

Pteronarcys californica (Newport) in Blacksmith Fork, Utah." *Aquatic Insects* 30: 187–85.

Pteronarcys californica

Life Cycle: Egg, nymph, adult.

Larvae: Known as nymphs. Salmonfly nymphs grow to approximately 3 inches long and are long lived; depending on water temperatures, nymphs will live for 3–5 years in the water (the colder the water, the slower the maturation) before emerging as adults in the spring or early summer.

Pupation: No pupation. Nymphs emerge from water as larvae and have a final molt to adult stage in a few hours.

Adults: Adults emerge together in a large mass in late spring to early summer. They are brown with distinct orange markings, up to 3 inches long. Adults live for a few days to 1 week.

Feeding: Nymphs feed on decomposing tree leaves. They do not feed as adults.

Habitat Indicators: Salmonflies live in high-quality streams with high oxygen concentrations and cool to cold water temperatures. They are typically associated with riparian debris leaf packs or on the underside of large cobbles. They are sensitive to pollution and seem particularly sensitive to herbicides and pesticides. Because of their large body size and big populations, they are an important food source for fish as nymphs, and, when they emerge as adults, for birds and bats.

Part III

Sleuthing for Caddis

About Caddisflies (Trichoptera)

The widespread, diverse caddisflies can be divided into three subgroups based on the kinds of cocoons and cases they build. All use silk, produced by silk glands, for at least part of their lives. They are closely related to moths; however, unlike aquatic moths that make water-soluble cocoons, caddisflies make cases with water-insoluble silk that can last for years after the larvae or pupae have emerged. The most primitive caddisflies make closed pupal cocoons, and some members of this subgroup are free living (without cases) as larvae. In the second subgroup, case-building larvae can make dome-shaped abodes that provide protection while they forage for food; others make fixed retreats incorporating silken nets. The most advanced subgroup of caddisflies build portable cases. They engineer cases with a wide range of plant and rocky materials in a host of shapes that include rectangular, boxy cases; sandy, tubular forms; sturdy, gravelly cones; and even spiraling, snail-shaped imitations. Cases are good protection against predators such as larger invertebrates, fish, or salamanders. They also increase the flow of water over the caddis's abdominal gills. Caddisflies without cases must depend on oxygen in the stream current and are limited to cool, moving water habitats.

Morphology

Caddisflies have well-developed heads with small eyes and short antennae. The thorax has three segments, each bearing a pair of crawling legs. The segment closest to the head (that is, most anterior), always bears a hard case made of cuticle on the top (dorsal) and bottom (ventral) surface. The second and third thoracic segments are variously covered (or not) by hard, cuticular plates and small buttons of cuticle, or setae. These characteristics help identify the various families, genera, and species of caddis. The abdomen bears tracheal gills along its length that are used for respiration. Small legs, called prolegs, at the hind end of the abdomen have sturdy claws that help the caddisfly move around; in free-living species, these prolegs are very flexible and mobile. For portable case builders, the hooks help the larva grip the interior of its case. They crawl by extending their thoracic legs out of the case.

Adult caddisflies are generally plainly colored, similar to moths. Their wings are covered in fine, hair-like microtrichia, like pre-adult subimagoes of mayflies. Caddisflies hold their wings in a tent-shaped form, and they are fairly good fliers, though they don't move great distances from water. Like the moths and butterflies to which they are distantly related, caddisflies consume pollen as adults.

Life History

In general, caddisflies in temperate zones have one generation per year (that is, they are univoltine), though this varies a bit between species. Depending on climate conditions, some caddis have life spans of one and a half or two years; other species might have two or three generations per year. After hatching from eggs, most larvae molt five times, undergo pupation, and emerge as adults that live in the terrestrial environment, often for several weeks. Diapause, when development is suspended for a time, occurs in many species, and can occur in egg, larval, or adult stages. This phase helps the insect avoid unfavorable conditions such as annual drought, reduced food resources, or extreme temperatures. For species that spend many months as larvae, this provides a mechanism for them to synchronize emergence so that adult males and females are present at the same time.

Use as Bioindicators

Caddisfly larvae are universally considered to be good indicators of water quality, particularly in streams. Both the lay public and professional biologists can easily recognize larvae. Sometimes the cases blend well with the habitat, but the trained eye can pick them out on the rocks, woody debris, and other surfaces.

Finding the larvae is usually taken as a sign for good habitat conditions. The diversity and abundance of caddis are used as measures of biological condition. However, some caddis, usually filter feeders, are more tolerant than other species of high nutrients; these species should be considered exceptions in evaluating water conditions because they can be misleading as bioindicators.

References

Wiggins, G. B. 1996. *Larvae of the North American Caddisfly Genera (Trichoptera).* Toronto: University of Toronto Press.

Wiggins, G. B., and D. C. Currie. 2008. "Trichoptera Families." In *An Introduction to the Aquatic Insects of North America*, edited by R. W. Merritt, K. W. Cummins and M. B. Berg, 439–511. Dubuque, IA: Kendall Hunt.

6 A Case that Glitters

Vincent Resh

Long, long ago Aristotle spotted insects crawling in streams, lumbering around with cases created from bits of wood: he called them stick worms. Unbeknownst to him they were actually the larvae of insects now called caddisflies, the adults of which look like moths. Centuries later, in the Middle Ages, the cases built of silk with a variety of materials attached reminded someone of "caddis men," who were that time's version of traveling salesmen, selling ribbons and cloth pieces from samples attached to their coats. The analogy was a good one, and the name "caddisfly" stuck.

The caddisfly *Neophylax rickeri* occurs in streams around the San Francisco Bay, near where I live, and is distributed north to British Columbia and Alberta. The genus name *Neophylax* refers to all of the insect's close relatives, but the species name *rickeri* is a patronym—a Latinized term fashioned after a person's name—in this case, in tribute to William Ricker. In 1935 Lorus Milne named this caddisfly species to honor his friend Bill Ricker, who collected a specimen of it in Culter Lake, British Columbia. Ricker, a Canadian scientist, was only twenty-seven years old when bestowed this honor. Later on, he became famous both as the "godfather" of fisheries science and as an aquatic entomologist noted for his work with another group of water-inhabiting insects, the stoneflies. But

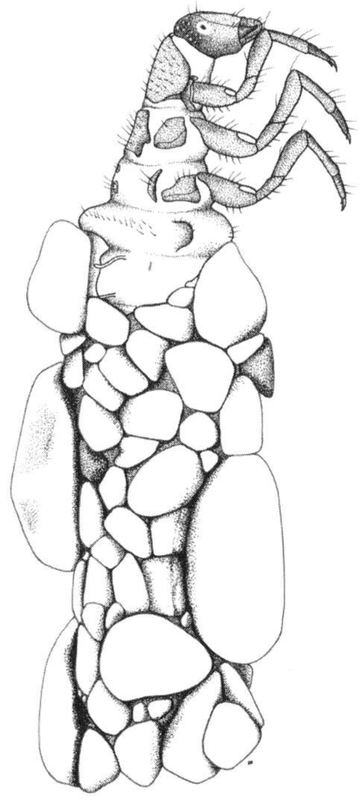

Neophylax rickeri in case

Bill Ricker was not the only person to have an insect named after him. Raymond Hays made Montana's Hyalitte Creek one of the best-studied stonefly streams in North America and co-authored a study of the stoneflies of Montana. In honor of his contributions, two stoneflies—*Isocapnia haysi* and *Nemoura haysi*—were named after him. (However, while Ricker and Hays were certainly worthy patronyms, others do not seem quite as "honorific": in 2005, a scientist named a slime mold "*vaderi*" after Darth Vader of Star Wars infamy.)

We can think of a species' scientific name as if it were a tag used in place of a more complex description that would require many more words. A scientific name also represents a hypothesis about the organism's relationship to other similar organisms. The name predicts that some of the information stored at the lower levels of taxonomic hierarchy (for example, the species name *rickeri*) may also be held in common with members of the more inclusive group (in this case, the genus *Neophylax*, the family Uenoidae, or the order of the caddisflies, the Trichoptera). When we find *Neophylax rickeri* in a stream we are reminded not only of its genealogy among caddisflies, but also our scientific ties to Bill Ricker.

Over thirty years ago, students in my laboratory and I found *Neophylax rickeri* in a creek near the Berkeley campus at one of California's state parks. We began a series of studies, which would be continued by many generations of students, to work out the general biology and intimate details of the *N. rickeri* life cycle. Early on, we realized "mother" *N. rickeri* caddisflies were pretty picky about where they placed their donut-shaped egg masses. Typically, there are many aggregations of egg masses on a rock, laid underwater on the downstream-facing sides and undersides of rocks. Certain rocks are chosen over and over again by different females and are covered with many egg masses. Because there could be from three hundred to over one thousand individual eggs, spherical or oval in shape, yellow to amber in color, all within a single egg mass, the total *potential* number of caddisflies (if all survived) could be enormous. However, only a very small percentage of eggs will become adults. Throughout the year, they might be eaten by fish or other predators, or they might perish from a myriad of environmental factors. But the eggs have some built-in mechanisms to help them succeed. For example, eggs in masses laid at the same time, or even within the same egg mass, don't develop at the same time. This has a bet-hedging advantage if a short-term catastrophe happens; without this mechanism, all eggs in a generation could be lost.

When we went hunting for egg masses, we found them at different times of the year, even among streams that were close to each other. In Northern California, we found eggs occurring variously in the fall, fall and spring, or fall through spring; these patterns depended largely upon which stream was being investigated. With my students, I wondered what advantage there might be to all the variation. We noticed a particular species of fly that laid its eggs on top of the caddisfly egg masses. To date, this tiny black fly (which we know is in the genus *Acanthonema* because of its morphological similarities to other known species of this group) has not been given an "official" species name; however, we have learned a lot about its biology. As soon as the small fly maggot hatches, it starts eating the individual caddisfly eggs. Many times, students have watched a single maggot devour most of the eggs within a mass. We came to think that *N. rickeri* females, in response to this unusual egg predator, might prevent losses to their local population by aggregating the egg masses together and laying them at different times of the year, depending on the presence of the predator.

Upon hatching, the *N. rickeri* larva gathers silt and fine sediment to begin making a case. These particles are attached to remnants of the egg mass with silk the larva releases from a gland opening just below its mouth. When they were picked up, these animals retreated into their case, and we only saw the tip of their heads peeking out. Nevertheless, the first thing anyone notices about most caddisflies are their cases, and the larvae of *N. rickeri,* as with most caddisflies, are the most readily recognizable life stage. The cases of different caddisfly species also vary with their age.

A *N. rickeri* larva develops through five spurts of growth (called instars), shedding its old skin with each spurt. To make room for its growing body, it enlarges its case by searching for a perfectly sized pebble, grasping it with the front pair of crawling legs, and attaching it with silk to the front of its case. Eventually it adds a series of slightly larger "ballast" stones on each side. These ballast stones have long been thought of as a way the larva holds the case down in turbulent flow; however, when we removed the ballast stones, we noticed that the caddis weren't washed away. So the reason for adding these stones on the side of the case is still unknown.

One year, a group of students wanted to watch case building in action, so we made a video showing this activity. Students dropped in glitter that turned out to be highly prized case-building material. Young larvae, newly hatched from the egg, quickly picked up pieces of glitter to make their cases. We were surprised to see larvae fight over single pieces of glitter and steal pieces of glitter from other cases! For some, there was a fight to the death. A young larva left without a case covering itself would soon die. Now we understood that availability of the right-sized particles for building cases might be a critical factor limiting caddisfly abundance and occurrence.

Fully-grown larvae begin something called a pre-pupal diapause within the sealed case built by the larva. Diapause simply means a pause or cessation in development—it's like hibernation in other animals. For *N. rickeri,* this activity (or really lack of activity) lasts from as early as March until late October. This is the way many insects avoid adverse seasonal conditions, such as winter in temperate zones, or hot, dry summers during periods of food shortage in the tropics. We've never been able to figure out why this particular species went into a diapause in the mild California spring and summer. But because other *Neophylax* throughout the world undergo a pre-pupal diapause, perhaps it's just a tendency that continues to express itself as a result of *N. rickeri's* shared evolutionary history.

The prepupal stage leads directly into the pupal period, a time of complete body rearrangement when the worm-like larva is transformed into the moth-like adult. After a couple of weeks spent in this pupal stage, the adults emerge by popping open the case, swimming to the surface, splitting open their transparent pupal skin, and resting to dry their wings. Adult life for *N. rickeri* is short: only a few days to a week. In a population of *N. rickeri*, males emerge earlier than do the females. So, as is often the case in human singles bars, the males are sitting around, waiting for the females to come in. To ensure offspring for the next generation, males of short-lived insect species *must* be available when the females are ready to mate. In the case of *N. rickeri*, soon there would a new generation picking their way through the stream, looking for perfectly sized pebbles for their cases.

It has taken generations of caddisflies and many cohorts of students to understand the annual patterns of *N. rickeri*. Along the way we've needed careful observations, ingenious experiments, and a strong sense of scientific curiosity. Like the caddisflies, we've sought the perfect pieces to put together the story of these intriguing insects.

Neophylax rickeri

Life cycle: Univoltine.

Larvae: Larvae hatch from large egg masses laid under rocks. There are 5 larval instars. Larvae build cases that are about 1/2 inch long at maturity, made of small pebbles. In the last instar, larger stones (ballast stones) are placed on each side.

Pupation: Prepupation up to 8 months (March to October). Pupation lasts about 2 weeks prior to emergence.

Adults: Pupae swim to surface, emerge as adults, and rest on vegetation or on the ground to dry. Males emerge before females, and adults survive from a few days to 1 week.

Habitat Indicators: *N. rickeri* occur in small, cool streams. Because they feed by grazing diatoms and algae on stream rocks, they require rocky substrates fairly clear of silty sediments.

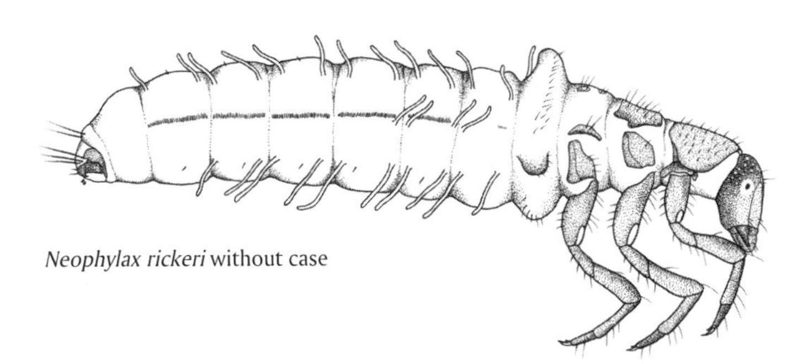

Neophylax rickeri without case

7 Life in a Cornucopia

John Richardson

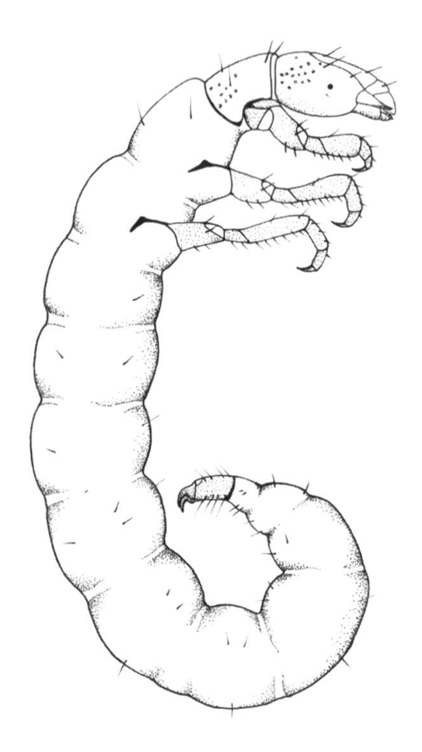

The first time I saw net-spinning caddisflies, I was staring at a damp layer of moss covering the surface of a rock in my professor's lab. It took me a while before I discovered little caterpillar-like beasties: they were larval caddisflies that were crawling slowly out of the moss. In the stream where we picked up the moss-laden rock, they had built silken nets. Our rude interruption of their normal habitat had forced them out of their nets.

All caddisfly larvae make silk to use in various ways, but net-spinners like the ones I saw in the moss have intriguing techniques for making silken nets to catch their food. In order to trap tiny particles of food flowing in the current, the larvae secure nets to stones, to rooted plants, or inside the crevices of logs, where the nets act like sieves to snag the prey. Some species make very fine nets attached to the bottoms of stones where flows are lower, while others live on the top surfaces in high current velocities. A few larvae make small excavations in fine sand bottoms of slow-moving streams and construct little "funnels" that stick up into the flow above the bottom.

Net-spinners represent several families of caddisflies, and many species are found in running water around the world. One net-spinning caddisfly family, sometimes called the "tube-maker caddisflies," consists entirely of predators. Their "tube-maker" moniker hardly does justice to the range of devices this group has devised to capture prey. Each species has a slightly different net design, but they are all very effective at trapping smaller animals as prey. The nets are dual-purposed, also providing hiding places from bigger predators.

When I began my graduate studies, I spent weeks and months dreaming up ideas for finding an ideal tube-maker to study and looking for streams to test. During my wanderings around the central Alberta countryside in search of good study sites, I came across a stream flowing out of a large lake, where I'd read that several species of stream insects were commonly found. While looking among the aquatic plants wafting with the current and searching along stones on the stream bottom, I found huge numbers of a tube-making caddisfly species with nets that looked like Thanksgiving cornucopias. My search for the right study caddisfly was complete.

These fascinating caddisflies were *Neureclipsis bimaculata.* New to me at the time, the species had been officially named by Swedish naturalist Karl Linnaeus in 1758. *N. bimaculata* is found across the northern tier of North America and Europe. The larva builds a net that looks like a cornucopia, ten times longer, or more, than the length of the animal itself. Once the larva reaches the end of its juvenile development, the net can be about fifteen centimeters long. While it is growing, the larva lives in the narrow end of the net, where a secure wrap of silk is often tucked very close to the stem of the plant to which the net is attached. A predator could easily eat right through the small tube at the end of the net, but the larva is well hidden. The open funnel of the net is attached on one side to the plant or stone and the rest of the net wafts with the flow of the water, displaying its wonderful engineering: it's not torn apart, despite how flimsy it appears.

I brought some *Neureclipsis* to my laboratory and placed them in aquaria to keep for a while. Within a few hours they had attached their nets to the aquatic plants and wooden dowels I'd "planted" for them. Just like a youngster feeding flies to spiders,

I tried placing various prey into their nets and was surprised by the unexpectedly exciting encounters. As soon as live prey hit the net, *Neureclipsis* rushed to the net; the unfortunate animal struggled very briefly as *Neureclipsis* hit its head into the prey, ramming it into the net, and spinning silk around it to immobilize the animal, just like a spider.

What trigger caused *Neureclipsis* to react? If a *Neureclipsis* larva sensed vibrations on its net as spiders do, then its response to dead prey settling onto the net might not be the same as its response to live, wiggling organisms. Sure enough, when I dropped dead water fleas (*Daphnia*) onto the webs, nothing happened. The live-capture strategy of the caddis appeared to work very well for capturing small, live lake zooplankton such as the water fleas or copepods that couldn't swim against even a modest flow. However, when we placed bigger, livelier stream invertebrates in the nets, the more mobile prey always managed to get out before *Neureclipsis* could get to them. No wonder we generally found *Neureclipsis* at lake outlets flowing into streams, places where it would be able to capture lake zooplankton aplenty.

It happens that many small zooplankton and other invertebrates have different activity patterns, moving up towards the surface, down into a lake's depths, or pulsing in their abundance in a stream. Would larvae grow differently if they caught food washing through the stream during the day as opposed to catching prey at night? My answer depended on feeding the larvae potential prey at different times of the day and night. To provide the right food, I also needed to collect zooplankton at the appropriate times. My approach was to sample near the road that ran over the stream draining the lake outlet; I threw my plankton net over the side of the bridge, and then filtered the sample to take back to the laboratory.

I'd been following this method for several days when, one night while throwing my net over the edge, I was blinded by the spotlight from a police car. It must have seemed an odd thing for someone to be throwing anything off a bridge in the middle of the night.

"Good evening officer," I said.

"What do you have there?" he asked.

I showed him my collecting bucket that already had one load of zooplankton, and described, perhaps in greater detail than he desired to know, how I planned to feed my caddisflies in the lab.

Thankfully, the officer was bemused. It must have seemed a story too crazy to be made up. "Okay then," he said, and left me to my solitary sampling.

Biologists are willing to go to great lengths, and some silliness, in order to answer our questions. After several more days and nights of collecting trips and hand feeding, we learned that the food *Neureclipsis* captured at night was more nutritious than what they caught in the daytime.

To emerge as an adult, *Neureclipsis* swims to the stream bank, still enclosed in its pupal skin. Awkwardly, it climbs up onto vegetation, and then bursts out of its pupal skin by squeezing its wings, legs, antennae, and body out of the cast-off skin. At first, it stretches, expanding its wings for a short time before it is fully able to fly. Males tend to emerge at a slightly smaller size than females and on average appear a week earlier. The love life of *Neureclipsis* can best be described as eager. The amorous males barely wait for the females to open their wings before they are courting them. That doesn't deter the females, who are clearly receptive.

Within a few days, females crawl down the stems of cattails or other plants, and attach their eggs to the plants underwater. In this species the tiny eggs take a very short time to mature. The diminutive youngest larvae measure less than one millimeter long (about the width of a sharpened pencil point). Even at this early stage of life, *Neureclipsis* build tiny nets that are hard to distinguish as individual nets, as they overlap each other on the plants and any other relatively solid surface they can attach to in the slowly moving flow. Still, they show the same amazing feeding behavior, albeit catching smaller prey.

Larval *Neureclipsis* build their webs on plants growing in the river bottom or even along the edge. If there is an old tree branch in the river, it will likely be covered with larval nets, where *Neureclipsis* wait in their tubular nets for food to arrive. Sometimes plants growing in the water can hinder water flow

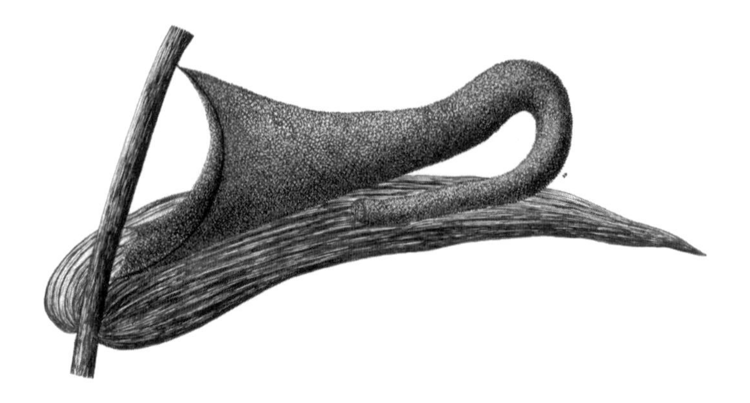

and in many places water management agencies mow the plants down to ensure efficient flows. For those reasons, to my horror, the leafy aquatic plants at my study stream draining Lac Ste. Anne were mown just as my study of *Neureclipsis* was ending. Despite frantic searching, few *Neureclipsis* could be found. I surmised that the larvae must have found new places to build their nets, because as the plants reestablished over the next week or two, new *Neureclipsis* nets seemed to blossom everywhere.

So next time you're near a slow-flowing lake outflow stream, search for the delicate-looking, cornucopia-shaped nets. They may be hidden among the grasses growing on the stream edges. Look closely: a clever predator may lurk within.

Neureclipsis bimaculata

Life Cycle: Univoltine; 4 stages of cycle are egg, larva, pupa, and adult.

Larvae: Larvae hatch from eggs and then start building a net. From egg to pupa requires 5 molts, as is true for most caddisfly larvae. Larvae stop feeding just before pupation when the last larval skin starts to detach.

Pupation: The last larval stage wraps up the end of its net into a snug sack of silk in which it begins to pupate. During metamorphosis, it cannot swim and is very vulnerable to anything that can get inside its pupal case. Pupation takes about a week to 10 days.

Adults: Pupae swim or crawl to emergent vegetation along the edges of the stream, and the adult splits out of the pupal skin. The newly emerged adult blows up its wings to spread them out, and then rests for about 10–20 minutes for the wings to dry before it can fly. On average, males emerge slightly smaller and a week earlier than females.

Feeding: Larvae feed on small invertebrates carried along in the water. They behave a lot like spiders, even spinning extra silk around prey to immobilize them.

Habitat Indicators: *Neureclipsis* larvae are relatively tolerant of warm water, but need good quality water. They are generally restricted to areas with relatively slow flow levels, mostly in lake outflow streams. As with many net-spinning caddisflies, they are sensitive to elevated levels of fine, suspended sediments that clog their nets, reduce the efficiency of catching food particles, can potentially damage the net, and dilute food quality of prey trapped on the net. Other net-spinning caddisflies show a range of tolerances to water quality.

8 Mystery of the Spine-Adorned Caddisfly

Marilyn Myers

Tucked into the arid Great Basin lands of eastern California and western Nevada, the desert springs I explored for my graduate research were full of surprises. Finding the springs proved to be only the first challenge. We knew scarcely anything about what lived in them; my goal was to discover and study the invertebrate diversity that lived in these tiny oases.

At first, I collected adult insects. There were the conventional approaches: setting out blacklight traps to draw insects in at night, placing emergence traps over the springs to catch adults as they emerged from the water, hanging sticky traps to catch adults on the fly, and sweeping with my wide net across plants growing next to the water. And there were the not-so-conventional techniques: putting out pans of water mixed with antifreeze to catch insects as they laid their eggs in my pans instead of the spring, and snatching by hand or by net the adult insects resting below undercut banks.

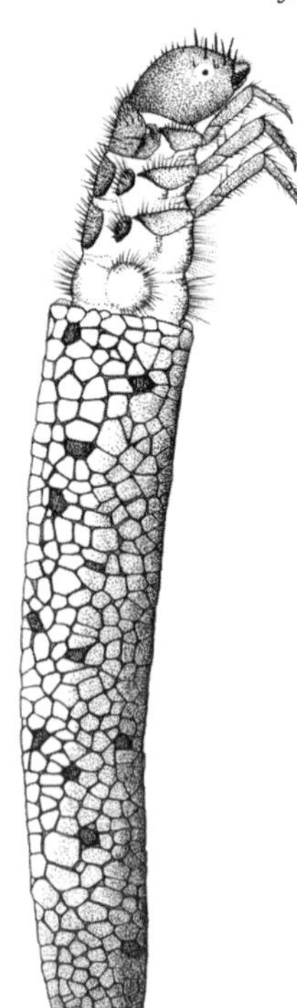

At a few of the springs, adults of the caddisfly *Pedomoecus sierra* were common. But the mystery was that my benthic collections rarely contained larvae of this species, and I had not collected any pupae. Adult *Pedomoecus* were small (only about a half inch long) and not known to be strong fliers. Because the desert springs were separated by miles of inhospitable habitat, I figured that the larval stages must be hiding somewhere in the spring runs.

Benthic sampling required innovation because the spring systems were so small. A Surber sampler, typically used for quantitative sampling, with its one-foot-square frame, was too big for many of the springs, and usually there was insufficient flow to carry the insects into the collection bag. Instead, I pushed a four-inch PVC pipe into the substrate and removed everything within the pipe. With these samples I could describe the animals I'd collected in a given volume and quantitatively compare across the many springs. For qualitative sampling I kicked samples into a D-frame or a small aquarium net. I sampled other microhabitats typically occupied by caddisfly larvae, like leaf packs, aquatic plants, and woody debris, by picking or scooping them up by hand.

Often I lay belly-down next to the spring run and just watched the flowing water and rocky bottom substrates, in the same manner a birdwatcher looks for movement in trees or shrubs. By watching quietly, I could easily see a large number of aquatic invertebrates going about their business in the crystal clear water, oblivious to my observation. Despite these efforts, my samples never contained more than an occasional *Pedomoecus* larva, and I never observed them on the rocky surfaces of the springs.

To help solve the mystery of the missing larvae, I went to the "encyclopedia" for trichopterists, *Larvae of the North American Caddisfly Genera (Trichoptera)* (Wiggins 1996), to find clues. Describing *Pedomoecus*, Wiggins wrote, "[t]he unusual morphological features suggest that the larvae have some specialized way of life, not yet understood." He was referring to an amazing assortment of stiff spines *Pedomoecus* larvae have on their head and legs: they appear as if they are armored for battle. In addition, the thorax and first two abdominal segments have girdles of long setae (hair-like structures) that give the larva a furry appearance. But the case or home of the larva, slightly curved and made of little sand particles, is unremarkable. It provided no clues about where the animal might reside. One day, despite conventional wisdom about caddisflies, I found larval *Pedomoecus* in the last place I expected.

The Glass Mountains of Mono County, California, were formed by volcanic forces, and most are composed of rhyolite, a rock that decomposes into angular, glass-like sand. After finding many adults at a small spring brook in those mountains, I knew the larvae had to be present, and I was determined to find them. Sitting bank-side along the spring, I observed discrete areas of sand that appeared as mini-dunes in the middle of the channel, constantly moving and changing. It suddenly dawned on me: that was one place I hadn't adequately sampled. Moving sand: no one would expect to find a caddisfly in that kind of microhabitat.

I grabbed a soil sieve, scooped up about a cup of sand, and gently swished it in the water. As the sand filtered through the sieve, about a half dozen *Pedomoecus* larvae were left stranded on the screen: the most I had ever collected at one time or place! More scoops of sand confirmed my sample site to be the specialized microhabitat of *Pedomoecus*. The strange spines and bristles now made sense: the larvae navigated through this malleable habitat using their spines and bristly legs, and the furry girdles around their abdomens prevented sand and silt from entering the spaces between body and case. Without the protective girdle, fine material could enter the case and interfere with its respiration. More searching revealed pupae, too. The last instar larva built a cap over the front end of the case and then the case was attached to a pebble, most likely to hold its place in the sand during the pupal stage.

Once I found the larval stages of *Pedomoecus* in the wild, I took some individuals back to the lab so that I could observe their behavior more closely. In the lab was a Plexiglas container constructed with walls that were only about an inch apart. I filled the container with sand collected from the spring, added water, and then placed the live larvae on the surface of the sand. Sure enough, within minutes they easily burrowed into the substrate. I watched them as they "swam" through the sand.

I had solved the mystery of where *Pedomoecus* larvae were living, and that discovery explained their unusual morphological features. In the process I learned that though a species may appear to be rare in benthic collections, collection of the adults may tell a different story. To accurately document biodiversity, it is important to collect both adults and immature stages of organisms. When we understand what animals require at all life stages, we will have a more complete picture of how they fit in an ecosystem—even ecosystems made of desert sands.

Pedomoecus sierra

Life Cycle: Univoltine.

Larvae: 5 instars, growing from 1/4 inch (instar I) to 5/8 inch (instar V).

Pupation: Occurs in sandy habitat and lasts 2–3 weeks.

Adults: Small, yellowish red; 1/2 inch long.

Feeding: Not known with certainty, but Pedomoecus may scrape diatoms off sand particles.

Habitat Indicators: *Pedomoecus* occupies small, cold-water mountain streams. Based on the specialized morphology of the larvae, pockets of loose, moving sand must be present as this is the unique habitat to which they are adapted. This microhabitat provides both protection (the larvae are buried) and food resources, while offering an area with little competition (few other invertebrates occupy this habitat). Most likely *Pedomoecus* is sensitive to pollution and water temperature, which may make the species susceptible to the effects of climate change.

9 Dicos for Ducks

Judith Li

Harlequin ducks are brightly colored birds that get their clownish name from their decorative white eye rings, white cheek spots, and burnt orange, white, and deep gray wing plumage. Our research team called them "harlys." In the winter of 1996, Kris Wright and I were into our second year of studying how harlys used the rock-cased caddisflies *Dicosmoecus gilvipes* for food at Quartzville Creek.

This alder-lined stream flows over boulder cascades and bedrock flats as it rushes through a canyon headed towards the Santiam River in western Oregon; during spring, it is a playground and nesting area for harlys. Adult birds spend the winter at

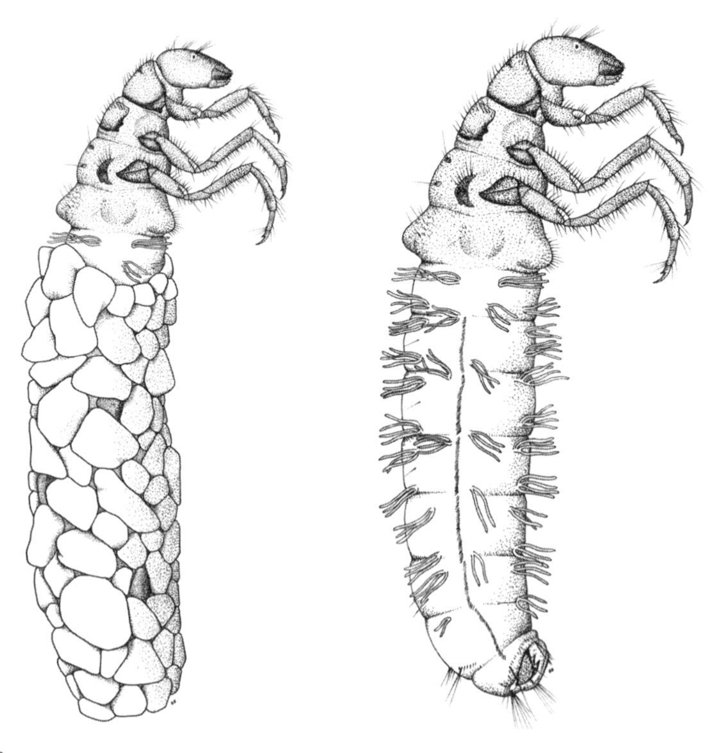

the ocean surf off the Oregon coastline then migrate inland to freshwater streams. Harly pairs build nests in the canyon's rocky nooks; the males then exit and females are left to raise their young. The chicks' gray and white markings camouflage them well, but in the spring of 1995, when we looked very carefully, we'd see chicks following their mothers, bobbing up and down in stream riffles, foraging for stream insects.

By late January of 1996, Quartzville Creek was running fast and full to the edge of the banks. Young *Dicosmoecus*, barely one half of an inch long, were crawling in the "surf zone" where the stream gently lapped against sands, gravels, and plant roots. The young caddis lumbered about with diminutive cases constructed of thin leaf bits and conifer needles. Each tiny larva's head stuck out as the front pair of legs crawled along, dragging the small woody case that protected its soft abdomen. The larvae made do with the pieces of flotsam available in early spring, and *Dicosmocus* is one of the few caddisflies out and about in early spring.

That year we expected to follow another annual cycle of caddis growth and harly activities, but things would prove to be different. On January 25th, the weather report was of a "rain on snow" event—rain and a heavy snowpack melting quickly during a sudden warm spell. That meant potential flooding downstream. In Corvallis, the Willamette River came about two inches short of cresting its banks and flowing into downtown. Kris heard that the stream runoff was very high in the mid-Cascades, so he set off on the hour's drive to Quartzville as soon as the highways were clear.

I waited for him in the warm comfort of our lab.

"Trees were down everywhere along the Quartzville road," Kris reported when he returned. "Before I could reach our study sites, I ran into a roadblock where the hillside slumped and had poured mud, boulders, wood across the road and into the stream—what a mess!"

Then Kris's mountain biking skills had kicked in. Climbing up and over the landslide, he had surveyed the damage very quickly. The view was quite a surprise: "The log the size of a semi-truck in the middle of our reach, the sandbar with the willows around it: they're gone! That huge boulder perched at the top of the little waterfall is nowhere to be found. From the signs of stream debris hung up in the tree limbs, I think the stream must have gone way up onto the banks where we set up our sampling last year. Who knows what happened to the critters in the stream?"

To answer that question we combed the stream looking for the caddisflies during March and into April. In normal years *Dicosmoecus* grow and continuously build cases using strands made by their silk glands, eventually including bits of sand. By late April they generally have molted three times, when their cases take on the shape of tiny cornucopia, skinny on the ventral, or tail, end, and about one quarter of an inch wide at the dorsal opening. This makes room for their heads and crawling legs. When their bigger, heavier cases are made entirely of sand and gravels, "dicos" have ballast to move out onto the bedrock in the middle of the stream. In Oregon, one-inch-long, rock-cased caddisflies grazing on algae in the middle of the stream are usually *Dicosmoecus*. But in spring of 1996, the year of the big flood, we saw only a handful of dicos.

Two months after the flood, right on schedule, fourteen harlequin ducks came back from their winter habitat on the coast—a normal number of males and females. Avian ecologists on our Quartzville study, Bob Jarvis and Howard Bruner, tracked the ducks' activities.

"Traditional nest sites are intact. Habitat seems sufficient," Howard told us. "They hung around, bobbed around in the stream for a couple of weeks, and then suddenly all but two females flew back to the ocean. What happened?"

Sitting under the alder canopy, our team gathered on the stream bank to consider the season's events. Kris and I reported what we knew of *Dicosmoecus's* life history at Quartzville Creek.

"Before this storm, the caddisflies were probably on a normal course. There was a long pupation in the stream under big rocks in the summer and adult emergence in the fall," Kris explained. "Remember, fishermen call them 'October caddis' because *Dicosmoecus* emerge in mid-fall. Pupae transformed into adults chew holes through their stony cases and swim to the water surface, just as salmon and steelhead return from the sea to spawn. They are about an inch long, like a fair-sized moth. Good fish food for trout and salmon. The lucky caddis that escape flutter their plain brown wings into this alder and conifer canopy overhead. That's probably where they feed on pollen for about a month before mating."

"The next part we know from our lab studies," I said. "By Thanksgiving, adults mate and lay eggs in bundles about the size of small marbles on the stream banks, in moss or other soft vegetation. They are counting on stream levels to rise during the winter and wet the eggs. About the time of New Year's Day, egg cases expand with a gelatinous matrix surrounding the tiny eggs, each about the size of a pin head. That is when larval development begins."

Kris jumped in, "It's all about timing. Most of the eggs had hatched and very young dicos were beginning to crawl about when the flood waters probably scoured away most of this year's offspring."

Howard and Bob had been studying declining population pockets of the playful harlequin ducks for several years. "You couldn't find the caddisflies in the stream, and the harlys had the same problem," Howard observed. "They must have decided they couldn't raise chicks this year."

The storm had rearranged the rocks and logs in the stream. Later, its consequences helped rearrange our ideas about harlequin ducks and caddisflies. Now we appreciated how the timing of abundant *Dicosmoecus* grazing in the middle of the stream was a critical coincidence for returning harlys who depended on the caddisflies to feed their young. Without the dicos, the returning harlys abandoned their nesting grounds after the storm, but we hoped they would return to Quartzville Creek the next year.

Dicosmoecus gilvipes

Life Cycle: Univoltine (sometimes semivoltine where winters are long and the growing seasons short).

Larvae: 5 instars, from 1/2 inch (instar I) to 1 inch (instar V).

Pupation: From 1 month (where fall temperatures are cool and early) to 2 months; prepupates where winters are moderate and snow infrequent.

Adults: Dull brown; at least 1 inch long; emerge and feed for about 1 month, generally in late fall.

Feeding: Larvae scrape algae; adults likely consume pollen.

Habitat Indicators: Fluctuations in flow, insect migrations, and bird foraging on these bedrock reaches in Oregon provide vivid images of habitat where we expect to see abundant *Dicosmoecus*. Gradually sloping banks where winter flows eddy out create microhabitat for very young larvae. Mid-sized streams with perennial summer flow provide habitat through the long growing season of this caddisfly's larval development. Sufficient light through an open canopy encourages the growth of algae that these herbivores depend upon. Large boulders provide surfaces for pupating larvae, and overhead canopy gives refuge for adults during their autumnal terrestrial phase.

10 Digging in a Ditch for Caddis

Norman H. Anderson

When our children were young, my family and I spent many weekends camping or picnicking on the Oregon coast or in the Cascade Range. We always chose sites along streams or lakes, where we looked for bugs, fish, and tadpoles. As an aquatic entomologist, I found these outings to be a great way for combining business with pleasure—I was actually being paid to go on picnics!

Quartzville Creek, in the western foothills of the Cascades, was one of our favorite destinations. The sparkling water tumbled over rocks and stones as it rushed through a canyon on its way to the lake behind Green Peter Dam. One early spring day in May 1973, while my wife and the children explored the stream, I crossed the road to see what I could find in a ditch that was fed by a hillside seep. At first glance there wasn't much of interest, but then I noticed activity on the surface of the mud and moss. Using a tea strainer as a weapon, I soon had a diverse fauna of swimming and crawling creatures in my collecting pan—mayflies, damselflies, beetles, and caddisflies. The most abundant specimens didn't move at all. They were the empty cases of caddisflies that I figured had just emerged. The half-inch-long cases were made of sand grains. One end of each case was attached to wood or stones by silken threads, and the opposite end was open where the caddis had chewed its way out.

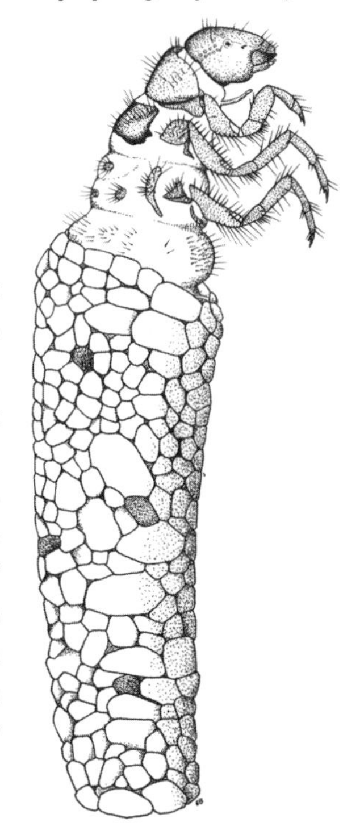

Turning over some pieces of bark beside the ditch, I found several egg masses—tough, gelatinous tubes about half an inch high. As these were likely deposited by females that had emerged from the empty cases, the chances were good that some adults might still be around. I went back to the car for my collecting net and, with a few sweeps in the streamside vegetation, caught several adult caddisflies. Back in the laboratory they were identified as *Pseudostenophylax edwardsi* (family Limnephilidae). That Sunday picnic led to a thirty-year study of this obscure species.

The following year I made monthly trips to the Quartzville ditch to trace the life cycle of *P. edwardsi*. The species' annual generation begins in April or May when adults emerge and fly into streamside vegetation. After a few days they mate, and females deposit their egg masses on wood or stones near water. When the eggs hatch in about two weeks, the tiny larvae crawl into the water before they build their first case.

The larvae develop through five growth periods (instars), making mineral cases that undergo sporadic construction to accommodate the larvae's growth. Using silk as glue, they add sand grains on the front ends of their cases, and, by cutting off the smaller posterior ends, they enlarge their cases. Most of the larvae's growth takes place during their final instar, which occurs in late summer and fall, when they eat leaves that have fallen into the stream. In early spring, larvae attach their cases to woody or rock substrates, close both ends with a silk mesh, and transform to pupae. After about a month, the pupa bites off the end of the case and wriggles out. It crawls or swims to the water's edge, casts off the pupal skin, and emerges as an adult.

This life cycle pattern made sense where there was water all summer, but I was puzzled by places where the ditch was dry for two to three months. All along the ditch, whether wet or dry in the summer, final-instar larvae proved to be common when I sampled after the fall rains began. How had they survived the drought, and would the caddis from the dry area be smaller or less fit after not feeding during the dry period? My monthly sampling showed that by early spring these larvae were the same size as those from the permanent flow area, and adults weighed the same in either habitat.

It only took a little digging to figure out how *P. edwardsi* adapted to the summer dryness. When a site began to dry up, the small larvae (second to third instar) burrowed down into the sand and stopped growing until the flow resumed. In the fall and winter, when the food supply of autumn leaves would be abundant, final-instar larvae could catch up in size with those that had been feeding all summer in permanent flow. These adaptations were key to surviving summer drought.

We were able to rear *P. edwardsi* through two generations per year in the laboratory. This led us to compare it with other caddis species as a candidate for bioassays of water quality, but some pond species turned out to be more suitable because they were more tolerant of warm temperatures. *P. edwardsi* was, however, a favorite in our lab studies of behavior, food preference, and growth under various temperature and photoperiod regimes.

The availability of larvae in a lab culture encouraged students to raise questions about caddisfly behavior that could be answered by observation or by simple experiments. One student used food coloring to show that undulations of the abdomen drew a stream of water through the case, making more dissolved oxygen available. She observed that the rate of flow increased at higher temperatures. Another student studied the timing of case building by switching the type of sand every few days. He demonstrated that sand grains were added to enlarge the front end of the case, but he wondered, how did the larvae remove the small ends of their cases? He arrived at a partial answer when he saw that larvae positioned themselves headfirst in their cases to cut off the end from the inside. Then, with a lucky observation, he caught one in the act of turning around without leaving its case: the larva simply tucked its head between its legs and wriggled its way down until it was completely turned around.

During an outreach program to elementary schools, *P. edwardsi* was a star performer. Children were fascinated by these little tubes of sand crawling about. When touched, the larva would quickly retreat into its case, but after a short time it would poke out its head and legs and resume exploring. The larvae provided a "hands-on" experience, even for the most squeamish or cautious kids, because the sand-grain case could be picked up with fingers or forceps without hurting the larva—and it didn't pinch or bite!

In 1992, we moved from inside the city limits of Corvallis to a nearby hillside property. The backyard was part of an oak woodland with a small summer-dry stream. This was an ideal spot for me to monitor the aquatic insect community (travel time and costs were an absolute minimum). And there I found my old friend *P. edwardsi* as the dominant caddisfly. For fifteen years I recorded the adult emergence of this caddis, including a wet period and two drought years when few, if any, adults emerged. But the numbers bounced back when the rains returned. *P. edwardsi* proved to be quite an opportunist. This little caddisfly, living inconspicuously in small headwaters, temporary streams, and ditches, may have the life-history adaptations to survive through droughts or other unpredictable adversities that may come with climate change.

And, to think—it all began with a spring picnic!

Pseudostenophylax edwardsi

Life Cycle: Univoltine (but bivoltine when reared at 15°C with a long-day photoperiod).

Eggs: Deposited in a gelatinous mass on wood or stones above the water line.

Larvae: 5 instars; newly emerged larvae enter the water before building mineral cases. Final instar is the over-wintering stage.

Prepupae/Pupae: Develop in sealed larval cases for 6–8 weeks.

Adults: Emerge in early spring and live for a few weeks; they are brown, about 15 mm long, and tend to be day-active.

Feeding: Larvae are shredders of leaf detritus; adults imbibe water and probably nectar.

Habitat Indicators: Field observations described in this paper provide the model for habitat requirements of this caddisfly. Larvae are found in cool headwater streams, mostly those with intermittent flow. They can burrow in sand or small gravel to avoid desiccation.

11 A Criminal "Case" Made with Caddisflies

John R. Wallace and Richard W. Merritt

Approximately twenty years ago, I was crawling around a small stream on South Mountain near Shippensburg University in south central Pennsylvania with my master's advisor Fred Howard, studying the life history of caddisflies. Little did I know that what I was learning would come in handy during a murder investigation in south central Michigan. I became interested in how aquatic insects could be used in forensic investigations while working on my PhD at Michigan State University with my major professor, Rich Merritt. My interests have pursued the unusual and somewhat bizarre direction of forensic entomology ever since.

The arrival time, development rates, and succession of terrestrial insects on many food resources is fairly predictable. Because of what we know about their life histories, physiology, and behavior, experts can analyze the colonization of wood consumed by termites, foodstuffs invaded by beetles or moth larvae, and dead bodies that attract a host of other insects. This makes insects very useful to forensic studies in estimating the interval of time since colonization. However, aquatic insects have not truly evolved to feed on carrion. For the most part, they utilize corpses as extensions of stream substrates. In fact, their presence on a corpse may, in many cases, be happenstance. On land, terrestrial insect colonization of human remains has been intensely studied, but using aquatic insects in these types of investigations is severely limited. We

know that the key for precisely estimating the time from when a body was submerged in water to the time of discovery (which we call the Postmortem Submersion Interval Estimate, or PMSI) is to understand events in the life history of the colonizing insect. I became especially interested in the phenology, or timing, of specific life history events for aquatic insects found on human remains in aqueous environments.

Our case in point: around the second week of June 2005, the unidentified remains of a male victim, partially buried in cement and placed in a duck decoy bag, were recovered from the Red Cedar River, a river we fished and canoed during my PhD program. When several large case-building caddisflies were collected at autopsy, Rich Merritt called me because of my expertise with this group of aquatic insects. My task was to determine how long the caddis might have been on the body.

The victim was reported missing on March 21st, and the remains were discovered on June 9th; however, it could not be determined when he died or when he was placed in the water. Without considering the aquatic insects on the remains, the forensic team estimated that his remains might have entered the stream any time from when he was last seen to when his body was discovered.

I identified the caddisflies found on the remains to be *Pycnopsyche guttifer* and *Pycnopsyche lepida,* belonging to the family Limnephilidae. *Pycnopsyche* hatch within a gelatinous substance that is deposited by adult females along stream banks close to the water's edge. This jelly-like matrix oozes, with the first instar hatchlings in tow, into the stream. During the first instar, larvae feed on the bacterial/fungal layer that forms on leaf surfaces. Hatchling larvae construct cases out of bits of debris, but they gradually incorporate small leaf discs cut out during the second through fourth instars. As they grow, their diets incorporate entire leaves that are skeletonized by their mandibles. Sometimes the cases are flat; others are triangular in cross section. Like their close terrestrial relatives in the order Lepidoptera (butterflies and moths), caddisflies produce silk that has many functions during their life spans. Interestingly, *Pycnopsyche* larvae switch their case materials and microhabitat as they enter the fifth instar. In this last stage, they either construct cases of sticks for slow-water habitats or build cases with stones to move towards fast water. These cases serve two functions: firstly, they act as ballast in the stream current to maintain the larvae's position in the stream, and secondly, they function as camouflage, blending with other woody debris or small stones in the stream. Thus the larvae are protected against both aquatic (fish) and terrestrial (avian) predators.

I learned during my master's degree research that *Pycnopsyche* in Pennsylvania switch their case materials in early- to mid-March and continue to feed through May. By June, large organic matter (particularly leaves) needed for food becomes unavailable, water temperatures increase, and stream predators such as fish are more active. By mid-June the larvae attach their cases to the substrate and shortly thereafter construct a mesh net across the opening of the case. This very specific life-history event terminates active movement in the stream and initiates diapause, when feeding stops and metabolism slows. Still protected by its case, the caddisfly eventually becomes a pupa, and then undergoes metamorphosis into an adult caddisfly. This sequence of events proved to be pivotal information for estimating a PMSI in this case.

We found that some of the larvae on the body had attached their stick cases to the remains, but they had not yet constructed mesh nets, nor had they initiated pupation. Armed with this information, we were able to narrow down the time when those particular larvae had colonized the remains, providing a PMSI estimate of late April to late May. Keep in mind that this was not an estimate of the time since death; rather it estimated the time period when caddisfly larvae could have colonized the remains and therefore when the body had entered the water. It is possible that the remains entered the water earlier, but this time frame was the most likely colonization period by these larvae. In this way, both *Pycnopsyche* and biologist sleuths helped police build a case for when the victim had been thrown into the river.

As the field of forensic entomology grows, actual case examples like this one may provide fodder for further experimental investigation. There is plenty of room to explore the role of aquatic insects in forensic investigations.

Pycnopsyche guttifer and Pycnopsyche lepida

Life Cycle: Univoltine. Unlike many aquatic insects, these species of caddisflies begin and end their life cycle on land.

Larvae: Eggs are laid in stream banks where first instars are born. *Pycnopsyche* have 5 larval instars, and construct cases of leaf disks for the first 4 instars, They switch to cases of sticks or stones (1 inch long) in the fifth and final instar.

Pupation: Active fifth instars attach their cases to a log or rock. Pupation begins by mid to late June, lasts for approximately 2–3 months.

Adults: By late August to September, adult caddisflies emerge by cutting openings in the silken mesh caps of their cases, then swimming vigorously to the stream edge, a rock, or exposed log. The 1-inch-long adults live only 3–4 weeks. *Pycnopsyche* adults have distinct odors that may allow for species recognition when locating suitable mates upstream. Adult emergence lasts until mid to late October and coincides with autumnal leaf fall.

Feeding: Detritivores, feeding on decaying, decomposing organic matter in streams, e.g., leaves, flower parts, and occasionally animal tissue. Their mouthparts are designed to shred this organic material into smaller bits. As "shredder" organisms, they not only convert terrestrial leaf material into their own biomass, but they also reduce terrestrial particle sizes for other organisms to filter or collect.

Habitat Indicators: Distribution extends across the northeastern and midwestern United States, similar to the distribution of deciduous forests. *Pycnopsyche* require decomposing organic materials during their larval development and reflect the availability of those resources in habitats where they are found.

A case report for this story appeared as "Caddisflies Assist with Homicide Case: Determining a Postmortem Submersion Interval Using Aquatic Insects" by J. R. Wallace, R. W. Merritt, R. Kimbirauskas, E. Benbow, and M. McIntosh in The Journal of Forensic Science, *January 2008, vol. 53, no. 1, pp. 219–21.*

PART IV

Truly Flies

About True Flies (Diptera)

True flies are a group with which everyone is familiar, in one way or another. Mosquitoes, black flies, crane flies, and midges (sometimes called gnats) are among the best known aquatic Diptera because they are sometimes irritating to humans, often very abundant, and occasionally economically important. All true flies require some sort of moisture, which can come from quite a variety of sources like the inside of a plant or animal; for the purposes of this book, we are interested in those that live in bodies of water, primarily flowing water. Diptera have adapted to just about every habitat imaginable, except for open ocean. To reach their extraordinary levels of success, flies have developed a great range of adaptations for eating, breathing, and moving in extreme conditions.

Morphology

True fly larvae are highly specialized. They do not have segmented legs, but some have one or more pairs of short prolegs; others have abdominal creeping warts or turbercles that help them get around. Dipteran larval heads are very small at best. There are two major divisions of aquatic flies, distinguished by those with fully exposed heads in contrast to those with highly reduced heads. Larvae with well-developed heads have stout, toothed mandibles that bite or chew their food, as opposed to the claw- or hook-like mandibles larvae with reduced heads use to pierce or slash their prey. The highly modified mouthparts of black fly larvae are fans that they extend into the water for filtering food particles. The thorax usually has three segments that sometimes have respiratory openings called spiracles. Along the abdomen there may be a wild array of nubby tubercles, creeping welts, setae, or pubescent hairs. At the hind (or posterior) end of many larvae are various numbers of spiracles that are sometimes adorned with spines, making the posterior end of the animal more elaborate than the opposite, anterior end, which might have a barely visible, reduced head.

One feature that distinguishes the adults of true flies is a pair of knob-like halteres behind the larger, single pair of wings. The halteres are remnants of the second pair of wings and are used for balancing the insect in flight. Though some families of true flies spend adulthood skimming the water surface, most do not remain in direct contact with water. However, most aquatic adult flies stay close to water before laying their eggs in an aquatic habitat, with the exception of those that require a blood meal (such as mosquitoes, horse flies, or black flies).

Life History

Generation time for true flies vary greatly. Some species have one generation (univoltine) or two (bivoltine) per year; in these groups, adults are usually present in the spring and summer. Others have many generations per year (multivoltine). True flies undergo complete metamorphosis, having four distinct life stages. Usually there is a brief egg stage (a few days to weeks), followed by three to four molts as larvae when most are free-living. Pupae may be free-swimming, attached to the bottom, burrowed in the substrate, or simply pupated within the last instar skin. They emerge as adults either at or below the surface, and live anywhere from a few days to several weeks or months.

Bioindicators

As diverse as true flies are in their adaptations to the aquatic environment, so are their levels of sensitivity or tolerance to pollution and habitat alteration. Most Diptera are considered to be relatively tolerant of polluted conditions. However, the blephariccrids and *Deuterophlebia* described in our book are among the exceptions to the notion of dipteran tolerances. As a general rule, the longer-lived (e.g., univoltine) species comprise those that are sensitive, whereas the shorter-lived (e.g., multivoltine) species, such as mosquitoes, are tolerant of a variety of water quality conditions.

References

Courtney, G. W., and R. W. Merritt. 2008. "Aquatic Diptera; Part One. Larvae of Aquatic Diptera." In *An Introduction to the Aquatic Insects of North America*, edited by R. W. Merritt, K. W. Cummins, and M. B. Berg, 687–90. Dubuque, IA: Kendall Hunt.

Merritt, R. W., and D. W. Webb. 2008. "Aquatic Diptera; Part Two. Pupae and Adults of Aquatic Diptera". In *An Introduction to the Aquatic Insects of North America*, edited by R. W. Merritt, K. W. Cummins, and M. B. Berg, 723–25. Dubuque, IA: Kendall Hunt.

12 Encounter with Arctic Black Flies

Donna Giberson

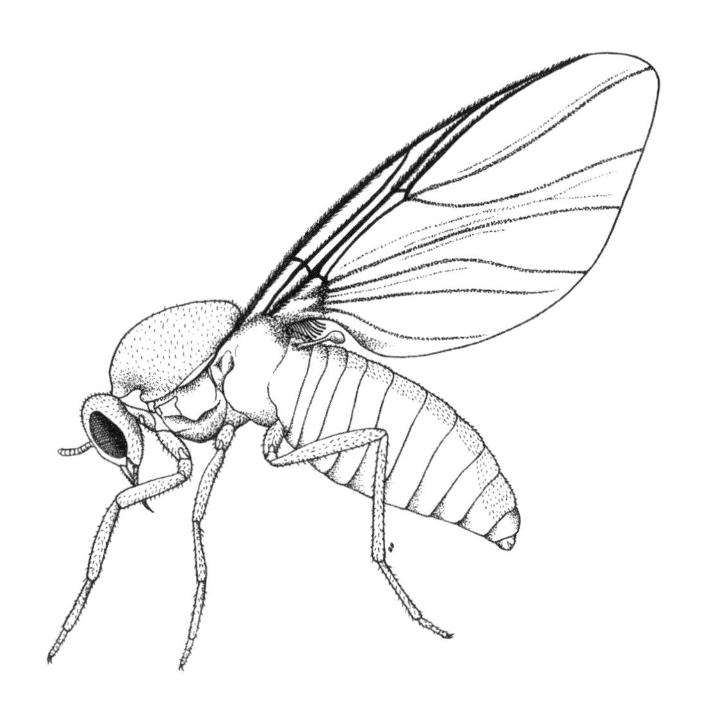

It was a beautiful evening in early July north of the Arctic Circle, with the sun dipping low but not quite reaching below the northwest horizon. Still adjusting to the wonder of the region's late-night sunshine, I surveyed my crew's heaps of field and camping gear as we began the task of setting up camp. The bright yellow Twin Otter that had dropped us into this wilderness disappeared over the horizon, reminding us that we'd be on our own for the coming weeks. Our crew of six had flown from Norman Wells with canoes and gear to survey aquatic insects in Canada's Northwest Territories. Far from any roads, we would spend the next four weeks paddling along the Horton River. Each of us had interests in particular aquatic insects, and we planned to sample for them as we made our way down a seven-hundred-and-fifty-kilometer stretch of the great river. The first night was spent sorting our collecting gear, personal stuff, and camping equipment, as we waxed poetic about the month to come.

Our method of sampling isolated arctic localities, which we called "collecting by canoe," gives a person a unique perspective on the countryside. We dropped in to remote regions by floatplane, paddled thirty to forty kilometers downstream each day, and sampled at various locations along our way. This particular trip involved a motley mix of mostly aquatic entomologists. I was sampling mayflies and stoneflies (Ephemeroptera and Plectoptera), Doug Currie and Peter Adler worked on black flies, specialist Brian Brown studied a group of terrestrial flies called phorids, and Mac Butler collected midges, those ubiquitous little flies that look a bit like mosquitoes, but don't bite. Our young river guide, Tim Gfeller, hired to keep us hale and hearty while we focused on our own work, rounded out the group. The Horton River is a fairly straightforward river to paddle for most of its length, though there are some interesting Class II and III rapids in the middle part of the river. These would lead to some exciting river moments, but luckily no major spills. The skill of our young guide was particularly appreciated, since it would allow us to run or line the rapids, and we didn't have to carry our heaps of gear around the dicey bits. (Packing for a four-week trip on an arctic river is enough of a challenge without also having to take full packs of sampling gear.)

The first day on the river was great: we made our way from Horton Lake to the river's main stem, which ran right along the northern tree line for the first few hundred kilometers. We exchanged greetings with some "fly-in" fishermen from a nearby wilderness lodge. "Might be the last people we see until the end of our trip," I thought. As the first day of paddling went well, so too did our sampling. I began to think, "This is going to be pretty easy, overall." Little did we know that just around the corner would be a scourge that would nearly scuttle our trip.

We were an experienced crew of aquatic biologists and river people who had faced just about every type of biting fly you could think of, so we weren't frightened by potential attacks by mosquitoes, black flies, and horse flies. We were equipped with repellents and nets—prepared, so we thought, for whatever the north could throw at us. I was facing a unique problem this trip, however, which made me the butt of many jokes during our trip planning sessions. After using DEET-based insect repellents consistently for years, I had developed a hypersensitivity to the chemical and couldn't use it any more. Unfortunately, most non-DEET alternatives of the time did little more than make one smell like lemon furniture polish (not good in tundra grizzly bear country), and they certainly were not effective against northern biting flies. The fellows laughed at my collection of head nets and bug shirts, but I was confident that my body armour would deflect the worst of the biting pests.

The next three days were arguably the worst of our lives. Suddenly, and with an appetite that is difficult to describe, hordes of black flies descended upon our little crew and took their pound(s) of flesh. Adult black flies feed by cutting into flesh then sucking up blood as it wells up to the surface. In seconds, we were all bleeding from numerous small wounds and doing all we could to block any exposed flesh from the depredations of the biting flies. Privately, we wondered how long we could tolerate this and whether it was time to call back the plane and withdraw with dignity. Publicly, we were stoic, refusing to give in while the others carried on. The DEET worked well for those that could use it, and so did the bug shirts; but at some point, we all had to expose unprotected flesh to the hungry blighters—when we were eating, or worse, performing those delicate bodily functions that often follow eating. Ironically, one photo taken during this three-day period became an extreme example of black fly bites in a medical textbook.

Peter, one of our two black fly specialists, helped us persevere in spirit, if perhaps not in body. He got us through those horrendous early days of the trip by pronouncing that this particularly vicious bunch of black flies was in a taxonomic group that had very specific habitat requirements; according to the map, Peter claimed, we would be out of their habitat within about three days. We clung to this information like a lifeline:

sure enough, within a few days we had paddled into different terrain, and the worst of it was over. We finished our trip, and though we had mosquitoes and black flies with us for most of the trip, they were easily handled in comparison to those early days. It turned out that one of the species that had bothered us so much was unknown to science. Peter and Doug were able to bestow an appropriate scientific name: *Simulium tormentor*, in dubious honour of the torment that we suffered early on in that memorable trip.

When making his prediction about the flies' habitat, Peter had made use of what he knew about the traits of immature black flies. Larval black flies live in flowing water, and most attach themselves to submerged rocks or plants, filtering (sieving) food particles from the water through fans on their heads. Each species is associated with a suite of environmental conditions, so it follows that if we know the species, we can predict what sort of habitat is present or needed for it to develop. Even though we had been immediately concerned with the adult stages, specifically with the females seeking blood for their egg development, adult distribution is closely related to larval distribution, so we could use one to indicate the other. Once Peter knew what species group we were dealing with, he could check the maps and assure us that we would be away from the group's habitat in a short period of time. The very specific habitat requirements of many aquatic insects are often used to monitor changes in habitat or environments, but in our case, it saved our sampling trip along an incredible northern river.

There were many moments on the river that made it special. We saw caribou for most of the way, but at one point, we looked up to see a large herd silhouetted on the top of the riverbank. Within minutes, caribou were pouring down the steep cliff like water from a tap, hundreds crossing in front of us and swimming through the river. After several minutes of back-paddling against the current so that we didn't crash into this new river obstruction, we were able to make our way to shore to observe this wonder without worrying about our canoes. When they had finally passed, our guide, Tim, gave a warning: "Those caribou head for the water to relieve themselves of hordes of biting flies, which rise off of them when they submerge themselves in the river. The thing to do now is to get back into the boats and paddle like hell to get through the cloud of flies!" Tim jumped into his canoe to put these words into action, but much to his surprise, he found us all in the center of the river, waving our insect nets and getting samples from that cloud of pestilence lingering over the river. What a chance for us! Many biting flies and other pests have specific hosts that they feed on. We'd gotten great samples of the ones that were attracted to us, but how often does one have a chance to sample the flies that feed upon a herd of caribou?

Another time, we had turned in for the night when we were awakened by a tremendous clatter. It sounded like a herd of something pounding right through our campsite. Peering out from our tents, we saw a single bull caribou galloping between our tents, pursued by an arctic wolf. Our campfire conversations for the next several days kept returning to this event, wondering aloud about what might have happened if the caribou had tripped on any of our tent wires and been caught by the wolf right in our camp.

The scientific results of that trip are still being analyzed and written up, but our crew was able to document many new species for the arctic, and we all gained a new appreciation of the linkages between the region and all the species that inhabit it. Would that other scientists could experience "collecting by canoe" as we did.

Simuliidae

Life Cycle: Black flies usually have 7 larval instars, and may be univoltine or multivoltine. Northern species typically are univoltine.

Larvae: Larvae develop in fast-flowing water. They are small and wormlike with well-developed head capsules and characteristic head fans for filtering food from the flowing water. They have one small proleg just under their "chins" and a row of hooks on the tips of their abdomens, which let them cling to a spun silken pad that they attach to rocks or vegetation.

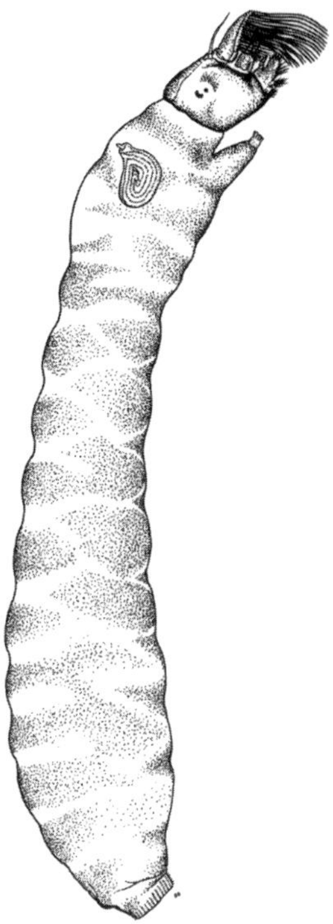

Pupae: Late-stage larvae select a pupation site and then spin a silken cocoon, which is attached to a substrate in the water flow. Pupal development may take anywhere from a few days to a few weeks, depending on water temperature.

Adults: Adults are small, dark flies, with hunched backs and wings that may be transparent or somewhat smoky in color. They usually emerge from their pupal skin in the morning and are active during the day. Adults of both sexes feed on plant nectar, but only the females suck blood to provide the needed nutrients to mature their eggs. Some species do not blood-feed, but most black flies do take blood for at least their first batch of developing eggs.

Habitat Indicators: Larvae develop in running water, and many species attach to rocks and vegetation that are exposed to the current. They can be found in the smallest spring seeps to very large rivers, though many species have more specific habitat requirements. Black flies feed by filtering tiny organic particles from the water, so they are most abundant downstream of sources of organic particles, like ponds, or places where livestock can access a stream. Most black fly species need well-oxygenated conditions, but some species can tolerate high levels of organic pollution. Adults are usually associated with the larval habitats, though they can be carried some distances from breeding areas by the wind.

13 Hanging on in the Alpine Tundra

Deb Finn

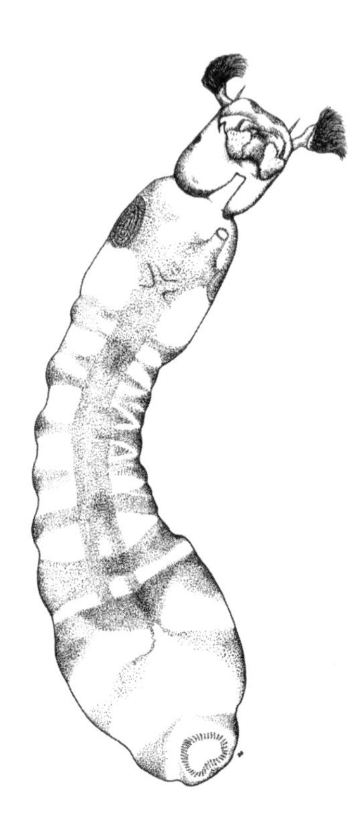

It was late July, but nearly freezing cold, the morning that my friend Kristy and I hiked to Chasm Lake in Rocky Mountain National Park. The lake occupies a deep glacial cirque that captures meltwater from Mills Glacier and neighboring snowfields. We were at an altitude of 11,760 feet, with Long's Peak soaring to more than fourteen thousand feet directly above us. Climbers busied themselves at the base of the glacier, loaded down with ropes, protective gear, and high-tech warm clothing, readying for a technical ascent of the vertical rock face known as "The Diamond."

We were similarly clothed for the cold alpine environment, but rather than ropes and quickdraws, we carried gear for collecting aquatic invertebrates. Kristy declared that it was well worth the extra layers of clothing; she was just happy to be away from the biting mosquitoes down below tree line. The cold, dry, often windy conditions in the cirque made for tough living conditions for many animals and plants, including biting insects. Vegetation was sparse, hugged close to the ground, and typically was hairy or cushion-like. This was not an easy place to live, although we did note a few bighorn sheep—naturally adapted to this rugged terrain—on the steep rocks above us. The harsh beauty of the crags, molded by the raw elements of nature, was formidable but breathtaking.

Ironically, it was an insect that motivated our hike that morning. We were looking for *Metacnephia coloradensis*, a strange and elusive black fly. At the time, there was only one known population in the North Boulder Creek watershed a few drainages to the south. We wondered about the fate of the species: if there was just one population remaining, would it be doomed to extinction if its habitat changed? Or could there be other populations in remote lake outlets that likely had never been explored? We targeted Chasm Lake that day because the Boulder Creek population flourished in the outlet stream of a similarly large, productive, high-elevation lake. If indeed other populations existed, it made sense that they would occupy a similar habitat type.

We skirted the lake, scrambling over boulders and bedrock toward the outlet stream. We could hear the rushing current long before we saw it.

"Oh no," I thought, "this is one of those 'buried' streams." "Buried" is the term I gave to streams that were free-flowing but covered in piles of loose boulders, too large to move aside. These streams were very difficult to sample.

We poked around, following our ears. Kristy noted the abundant spiders that spun large, durable webs among low-lying boulders. We found them only where the sound of the stream was loud and in crevices that seemed protected from the wind. The spiders cleverly placed their webs above the stream to capture valuable protein in the form of emerging insects. Small midges and even caddisflies were trapped in several webs.

About twenty feet downstream of the lake, Kristy suddenly lay completely prostrate on a boulder, apologized to a spider, disengaged its web, and stretched her hand down between boulders and into the water below. "Oh, bummer!" she yelled, "I can reach the bottom but I can't get any of the rocks loose!" She knew that *M. coloradensis*, like many other black fly larvae, attach themselves to stable rocks directly exposed to the stream flow. Like the spiders above them, they produce silk. The larvae cling firmly with abdominal hooks to a small silk pad attached to a rock, filtering food particles from the current with specialized fan-like mouthparts. Kristy knew that the easiest way to collect these larvae was simply to remove rocks from the stream. Any attached black flies would come right along with them.

Kristy struggled for a minute or two, trying to pry a cobble free from the streambed. "Yeee-oowww, it's cold!" she cried, finally withdrawing a numb hand, sans cobble. However, I jumped for joy at the sight of her poor hand: its surface was covered with at least fifty black fly larvae, with some trailing behind, swinging up from the stream on small ropes of silk. These larvae were obviously very abundant in this habitat, like the *M. coloradensis* that had been previously observed at North Boulder Creek. In addition, the larvae on Kristy's hand bordered on an entire centimeter in length (quite large for a black fly), and *M. coloradensis* grow larger than other black fly larvae in the region. I was sure we had found what we were looking for.

I scrambled for forceps and a collection vial from my pack. "Hurry!" Kristy demanded. "My hand is freezing!" The alpine wind was not helping her situation, so I grabbed a few of the larvae as quickly as possible. Then Kristy warmed her hand while I took my turn reaching into the cold stream.

A few feet upstream, I managed to extract a rounded rock about the size of a softball. The surface that had been exposed to the current was completely covered in large writhing larvae clambering among and over one another.

"It looks like Medusa!" Kristy laughed, referring to the snake-haired legend of Greek myth. We forgot about our cold hands and took a few photos to record our triumph, just as psyched as the climbers now reaching the summit of Long's Peak.

A couple of years later, I convinced my friend Jeremy, an accomplished underwater photographer, to join me on another late July *M. coloradensis* trip, this time to the North Boulder Creek population. Jeremy hauled a massive pack filled with both photography and snorkeling gear (among other items) up and over a two-hundred-and-fifty-foot waterfall as we followed the stream through the alpine tundra to the uppermost lake in the basin.

When we reached this more "unburied" lake outlet, we did not have to search for long. "Aha, the Medusa effect!" Jeremy noted with a twinkle in his eye. Indeed, there on the submerged rocks were the densely packed *M. coloradensis* larvae. Upon careful observation, the stream bottom just below the lake outlet appeared striped in places, with alternating columns of brown larvae and dark gray pupae. What a sight!

"What are they doing way up here in this completely unlikely place?" wondered Jeremy, suiting up in heavy fleece and drysuit for a submerged black fly photo shoot on the cold alpine morning.

"These critters are a perfect fit for the conditions," I told him. "An alpine lake outlet provides the very cold water temperatures as well as the small food particles that *M. coloradensis* apparently requires to persist. When these requirements are met, this species is highly successful and excludes just about every other animal there." I looked at our surroundings for a moment. "In contrast to the harsh terrestrial environment above tree line, this lake outlet stream is quite stable, even pleasant. The lake moderates changes in streamflow—like floods from spring snowmelt or drying after late-summer droughts when much of the snow is gone. Plus, the food supply from the lakes is huge compared to that in streams without lakes."

"But what about adult flies that emerge from the stream?" asked Jeremy. "How do they deal with the cold, windy tundra?"

"*Metacnephia coloradensis* is not your typical black fly," I said. "In fact, it's almost more like a mayfly. It doesn't bite, and it has a very short adult life. The adults apparently avoid the wind and cold by mating on the sun-warmed streamside rocks, and they quickly lay eggs at the stream edges before their time is up. No one has ever even seen them fly!" I knew that these characteristics, along with the specialized habitat they require, helped explain the rarity of the species, as well as its probable sensitivity to changes in the environment.

Jeremy took around three hundred pictures that day, fully immersed and facedown in the cold alpine lake outlet. We went to great lengths to document this rare alpine insect, simply thrilled with its unique contribution to the diversity of life. However, further exploration has ended only in frustration. After a thorough search in the greater Rocky Mountain National Park region, we have identified only three other lake outlet populations, including the more elusive one at Chasm Lake. As temperatures warm with global climate change, we worry about changes to the restricted habitats of *M. coloradensis* and about the species' survival as residents in the spectacular Rocky Mountain alpine environment.

Metacnephia coloradensis

Life Cycle: Univoltine: adults emerge from alpine streams in August–September. Those nearer to a lake outlet develop faster and emerge earlier.

Larvae: Large: up to 1 cm in length, brown and "wormy," produce silk. Most black flies have 7 instars, but that has not been evaluated for this species.

Pupation: 2–4 weeks; pupae are dark gray in color.

Adults: Short-lived (1 to a few days). These are poor fliers who mate on the ground shortly after emerging. Females produce 80–100 large eggs.

Feeding: Larvae filter small food particles that often drift from a lake; adults have not been observed to feed, and mouthparts are incapable of acquiring blood in a typical black fly manner.

Habitat Indicators: Large, productive, lake outlet streams well above the treeline provide the quintessential habitat for *M. coloradensis*. Ideally, the stream bottoms are rocky, with loose organization of cobbles and boulders and lots of space exposed to the current, providing abundant food for the filter-feeding larvae. Cold water is essential and temperature probably cannot exceed a maximum of about 10°C. Sunny bedrock and boulders at the streamside provide places for mating activities, as well as for observers (typically human) to perch while watching or collecting these fascinating critters. Endemic (native and exclusive) to the Colorado Rocky Mountains.

14 Making the Case for an Aquatic Insect and Its Habitat

Richard W. Merritt

I teach a course called "Aquatic Insects" at Michigan State University. The course consists of both lecture and laboratory, with a collection required by students in the class as part of their laboratory grade. They examine specimens from our teaching collection during each lab, and the students are then required to go out to different bodies of water throughout the state and collect aquatic insects, which they preserve and identify. This collection that they put together is then turned in at the end of the course, and they receive a given number of points and subsequent grade on the collection based on the total number and kind of insects collected from different sites.

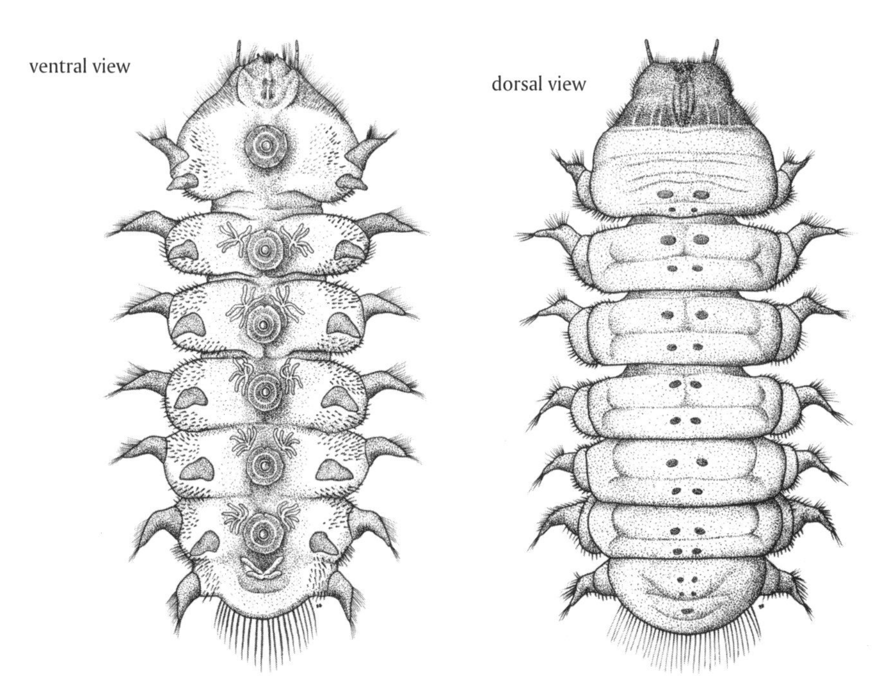

When I was grading the insect collections of the students at the end of the class one semester, I came across a larval specimen of the dipteran family Blephericeridae (net-winged midges) that one student purportedly collected in a stream near Michigan State University in East Lansing. I found this curious because the species had never been reported from the lower peninsula of Michigan. What a find!

I called the student in (for the purposes of this story, I will call her "Stephanie"), to discuss her find. "Stephanie," I said, "tell me about this particular insect in your collection. I would like to know more about its location and habitat."

"Oh, this one?" she replied, "In a nearby town, I found the larva in a stream that ran through the center of the town and through a park." She looked at me as if she had a question of her own.

"Is it possible that you mislabeled the location?"

Stephanie frowned, "No, I am pretty sure my location is correct. Why do you ask?"

"I am curious as to the habitat and would like to see your stream. Would you take John (my teaching assistant) and me to this stream?" Many thoughts ran through my head. If we could confirm the specimen's presence, it would be the first record of this insect in the lower peninsula of Michigan and perhaps would be worth writing up a short publication; it might also mean that I could possibly get a Nature Conservancy grant for studying a little-known insect in the lower peninsula of Michigan. However, I was a bit skeptical—Stephanie was not a stellar student, and her stumbling upon a find as significant as this did not seem probable.

The next day Stephanie, John, and I proceeded to the stream, which was a warm-water drainage ditch that meandered through the town and park adjacent to a baseball diamond. It was a muddy, silty creek at best, with no riffles or hard substrates of any kind for attachment by a scraper like the larva of a Blephericeridae. It was immediately apparent that this student had not made the connection between the habitat and the insect. She proceeded to point out three different sites in this creek, even though we kept saying that this was not the right kind of habitat for this insect to occur in nature. The condition of her sites did not improve as we walked along the creek.

At this point, I finally confronted her. "Do you know that this family has never been collected from the lower peninsula of Michigan, and it does not occur in highly sedimented streams like the one you are showing us?"

Stephanie finally broke down and stammered, "Okay, I admit that I didn't find it here." She looked uneasy. "I took the insect from your teaching collection to add to mine—I couldn't resist—I wanted to get enough points to get a passing grade. I made up a false habitat label. I didn't think you would notice. I am sorry."

I have thought about this incident from time to time. The bottom line was that this insect, its body form, and feeding behavior dictated the type of habitat that it was found in. Stephanie thought she could fool the experts: not so! The specimens in my teaching collection were from mountain streams in western Canada that contained habitats typical for this insect. Stephanie did not receive a passing grade for her collection, needless to say, and I doubt if she ever pursued a career in the field of aquatic entomology! But every time I remember this story I'm reminded that when we find aquatic insects in a stream, the very specific relationships between the insects and the environments in which they live help us make a pretty good case for the condition of their habitat.

Blephericeridae

Life History: Net-winged midges produce 1 generation per year. Adult females lay eggs on exposed rock surfaces during the dry, warm season. Larvae hatch from the eggs once they have become submerged by the rising waters of the rainy season.

Larvae: This stage of the insect is peculiar in that it uses ventral suckers (suctorial disks) to maintain its position on rocky substrates in torrential streams. A hydraulic, piston-like apparatus gives the larvae the ability to generate suction that allows their suckers to work—even waterfall habitats are occupied by the larvae of these net-winged midges.

Pupae: Pupae are firmly attached to rocks within the flow with permanent adhesive pads.

Adult: Emerging adults split their pupal cases by applying downward pressure against the substratum; adults usually reach the stream surface in air bubbles. Because the wings develop to full size within the pupal case and merely unfold during emergence, adults can fly immediately upon reaching the water surface.

Feeding: Larvae graze thin algal films (mainly diatoms, but they may also feed upon bacteria and fungi) off rocks within fast-flowing environments. Adult females with well-developed mandibles usually are insect predators, whereas males and non-mandibulate females appear to feed on nectar.

Habitat Indicators: The larvae of this family of Diptera are generally indicators of clean water as they occur in clean, highly oxygenated, fast-moving streams and rivers. In these habitats, the larvae attach to hard substrates like gravel, cobbles, or boulders, using modified suctorial discs attached to the ventral (under) side of their body. They use these suckers as holdfast organs to resist being swept away by the current. The larvae of these insects feed by scraping algae off the rocks and are always found in riffle or sometimes waterfall areas, and often in mountainous streams.

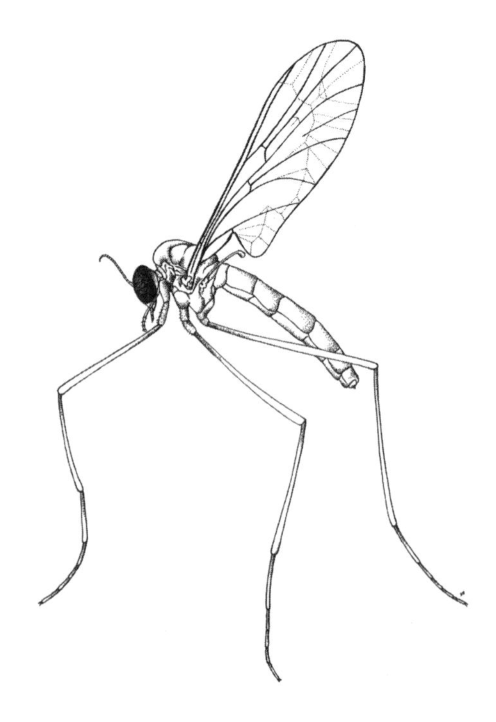

15 Way Cool Mountain Midges

Gregory W. Courtney

"Deuterophlebiids (du-ter-o-fle-bee-ids) are set up as a demonstration at the front table," read an announcement from Dr. Norm Anderson in our aquatic entomology lab at Oregon State. "These are rare flies, so be very careful when you handle the specimens."

He was certainly making a fuss over these flies called "mountain midges." This was the first time he'd placed a demonstration up front, next to his desk. Perhaps he wanted to keep an eye on the specimens? Keep an eye on us? Make sure we didn't damage those precious specimens, or walk off with one or two for our own collections (bonus points!)?

I had seen drawings before but could not have anticipated how exciting it was to see actual specimens of larval deuterophlebiids.

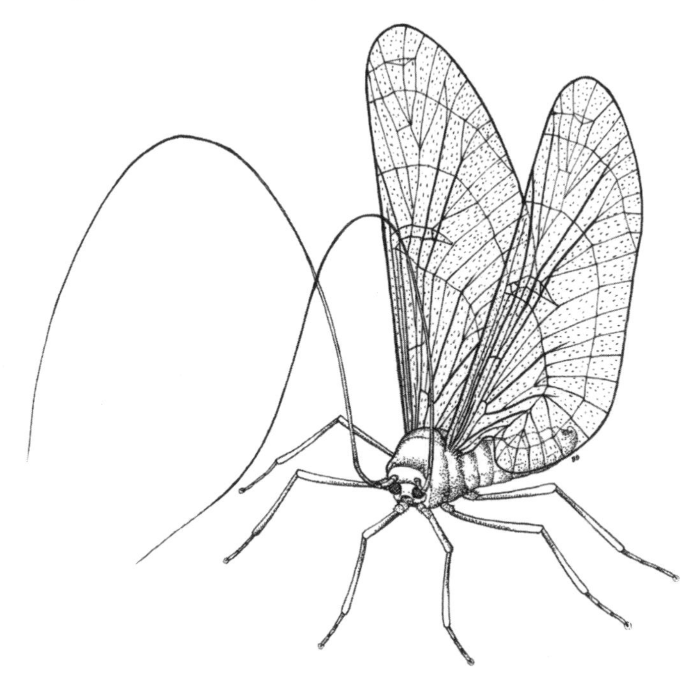

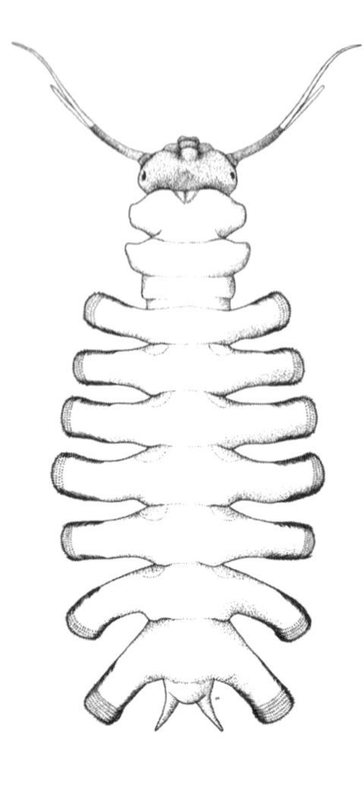

They were odd indeed! The body was dorsoventrally flattened, as though someone had stepped on the rock that harbored those larvae. Each of the first seven abdominal segments had a pair of large, lobe-like structures that could be turned inside out, and each of these prolegs, or "false legs," had several rows of hooks (called apical crochets). Way cool! The head, with those elongate forked antennae, had odd, comb-like mandibles and other weird mouthparts with enlarged spatulate (spoon-shaped) hairs. For a moment I wondered, "This can't be a real insect; is it a trick by Dr. Anderson?" I soon dismissed that idea. Nevertheless, I believed those larvae certainly looked more like something from a sci-fi movie than from some river in Oregon.

"Where were these collected?" I asked.

"The Marys River," replied Dr. Anderson. That was the river that ran out of the Oregon Coast Range into the Willamette River a short bicycle ride away.

"No, where *exactly* were they collected? When? Are these specimens all you have? What is the purpose of . . ." and here I would inquire about any number of the morphological features I might be considering. Dr. Anderson had come to expect a barrage of questions from me (perhaps he began regretting that the demonstration was so close to where he sat). He knew me well enough to be able to see the path ahead. I had already volunteered in his lab, where I'd shown some competence and interest in identifying Diptera (fly) larvae. At the time I hadn't fully realized that nobody else in the lab, especially Dr. Anderson, wanted to identify "maggots," so the niche was wide open (and my filling it would be encouraged). Besides, he knew that anything he called "rare" would tweak my interest. So, the combination of "odd," "rare," and "Diptera" meant the inevitable: I had to learn more about these flies, and of course collect some! Although it seemed a rather onerous task at the time, collecting has become less of a challenge after nearly thirty years of studying deuterophlebiids.

We now know that mountain midges are widespread in western North America, found throughout the Sierra, Coast, and Cascade ranges, and in the Rocky Mountains from New Mexico to the Yukon. However, when I was a student in Dr. Anderson's course, these flies had been collected from only a handful of locations. As it turned out, the specimens from the Marys River were collected just west of Philomath, Oregon, a mere seven miles from campus. This was the obvious destination for my next collecting trip.

So, less than two days after that fateful laboratory introduction, I was at the Marys, picking up rocks, taking kick samples, even "washing" rocks into my kick net. Although I found lots of interesting aquatic insects, there wasn't a single deuterophlebiid. I was crushed. Might these flies be as rare as Dr. Anderson suggested? The river was a bit high, so perhaps I simply hadn't been able to sample the best habitat. Regardless, it was back to the lab with the disappointing news. I asked him if there were other places in Oregon where these flies occurred.

"I've heard that Mike Stansbury, an aquatic biologist and avid insect collector, found them in the South Umpqua River. Part of an environmental assessment study," he told me. The South Umpqua drains out of the southern Cascades in a warmer, more southerly part of Oregon, about two and a half hours away.

"Time for a road trip!" I decided.

About a month after my failed attempt at the Marys, I was standing thigh-deep in the South Umpqua River. The river was moving along at one to two meters per second as I reached down into the chilly water and began pulling out rocks a little larger than basketballs. Paydirt! I found a handful of pupae at this site and at another, then both larvae and pupae at a third site later in the same day. Over the next couple of years I collected deuterophlebiids from nearly one hundred locations, and stepped onto a path of discovery, mystery, and graduate study of these unusual flies.

During this period I visited spectacular places like the East River in Colorado's Gunnison Basin, where I watched swarms of *Deuterophlebia* with tens of thousands or even millions of individuals. I didn't see separate masses of individuals over patches of whitewater as we often observe in emerging insects. Instead, it looked like a white mist extending up and down the river as far as the eye could see. Incredibly, this "mist" of deuterophlebiids was visible from the highway that paralleled the river! Rare flies? Not on that day or in that place!

Many classrooms and field trips later, and after a role reversal to that of a teacher, I had returned after a cross-country trip from the Midwest and was preparing for an early morning trip to the South Alsea River with my student Andy. Andy had a thirst for knowledge about flies, insects, and anything aquatic, but he was basically a night owl. "Dr. Courtney, why do we have to get up so early?" he asked. But I knew the question was based more on curiosity than an aversion to the hour. He was just as anxious as me to get to the South Alsea River.

"Because *Deuterophlebia* adults start emerging shortly after sunrise and the entire flight period is less than two hours," I said, hurrying him along. I knew we were already pushing the time of day and perhaps, the season.

We arrived at the river just in time to see a few males flittering back and forth over one of the riffles, but it was obvious we wouldn't see any large swarms like the ones I'd spotted in Colorado. Our best hope would be to net a few males and check some rocks for larvae, pupae, and maybe a female laying eggs. Andy and I quickly made a few sweeps over a stretch of whitewater.

"Got some!" I yelled to Andy.

He ran over just as I opened the net to peer inside. A handful of adult males were pressed up against the mesh, most flailing their wings or seemingly stuck to the net. "Why aren't they crawling up the mesh like most flies?" Andy asked.

"They can't walk. They have no tarsal claws to grab onto the net or other substrates; all they have are large pad-like lobes with a bunch of spoon-shaped hairs."

"What are the hairs for?"

"Good question," I said, "and the conservative answer is: we're not sure. Males spend their entire adult lives (only thirty to forty minutes!) flying over rapids, and they often get knocked onto the water surface. But they don't usually get stuck and often resume flight quickly. My theory is that their specialized legs act like 'snowshoes' to keep the fly from getting stuck on the surface film. Can you imagine how the modified tarsi, spatulate hairs on the lower legs, and maybe even the elongate antennae—four times the length of the body in some species—could help them?" I hesitated, then added, "You might see a few floating on the surface, but only at the end of the flight period, when males start to die."

Already we'd seen a few dying individuals, so we had to move quickly. Several more sweeps provided only about a dozen additional males; it was time to shift strategies and search for larvae and pupae. "Pick up rocks from the swiftest parts of the river, if you can," I advised Andy.

Soon Andy pulled up a rock that had an odd-looking larva attached to the exposed upper surface. The larva's movement was distinct, rotating its head and thorax like a pendulum as it progressed forward. "Is this one?" he yelled.

Sure enough. Once Andy knew what to look for, he saw an additional three or four larvae on the same rock. I pointed to several small, dark, flat, disc-like objects, each with a pair of crooked anterior horns. "And these are the pupae," I explained.

With a few more rocks, we soon had a good series of larvae and pupae. The only thing we hadn't collected was a female. I told Andy it was doubtful we'd find any flying: "Most will have already mated, landed on emergent rocks, and shed their wings when they crawled into the water in search of sites where they could deposit eggs." Still, I figured we had one last chance. "Andy," I directed, "start looking at spider webs."

The long-jawed orb weaver, *Tetrognatha elongata*, is far better than humans at collecting mountain midges, partly because their webs are constructed right over the stream—these are ideal locations for capturing aquatic insects. A lesson learned early in my studies of *Deuterophlebia* was that *Tetrognatha* webs were one of the best ways to collect adults, especially at sites that could be visited only later in the day (i.e., after the emergence period). Because of many overhanging branches, snags, and midstream debris jams, the South Alsea has always been an ideal place for *Tetrognatha* webs. But, alas, a successful collection "sweep" of *Deuterophlebia* life stages was not to be that morning. Although several webs contained entangled males, some still "kicking," we failed to find any females. This was not too surprising given that females have regular tarsi and are better at escaping sticky spider traps.

However, despite our failure to collect females, our two hours at the South Alsea was time well spent. And it was good to know that, after a lifetime of chasing *Deuterophlebia*, the resident population and the river were still healthy.

Oh, and Dr. Anderson no longer calls these flies "rare." "Rarely collected," he says instead. Fair enough.

Deuterophlebia spp.

Life Cycle: Univoltine; though some populations might have 2 generations (bivoltine) or multiple generations (multivoltine) a year. Active mostly in spring to early summer. Eggs are the overwintering stage. In populations from warmer rivers, this is often preceded by an egg diapause.

Larvae: Four instars, from 1 mm (instar I) to 3–5 mm (instar IV), all with forked antennae and crochet-tipped lateral prolegs.

Pupation: Asynchronous (that is, not all individuals enter pupation at the same time) in most populations. Lasting 1–2 months; pupal stage lasts 2–5 weeks (longer where temperatures are cool).

Adults: Dark with fan-shaped wings; males with long antennae and pad-like tarsi; emerge early in morning.

Feeding: Larvae graze algae; adults do not feed and live less than 2 hours.

Habitat Indicators: *Deuterophlebia* occur in cool, torrential streams of western North America and eastern and central Asia. The larvae and pupae inhabit the exposed surfaces of rocks, often where current velocity exceeds 2 meters per second. These habitats provide nutritious food (mostly algal diatoms) for larvae and abundant oxygen for larvae and pupae. These life stages appear to be especially sensitive to environmental perturbations, particularly increasing temperature and sedimentation.

16 Marine Sea Stars, Nudibranchs, and Midges

David Wartinbee

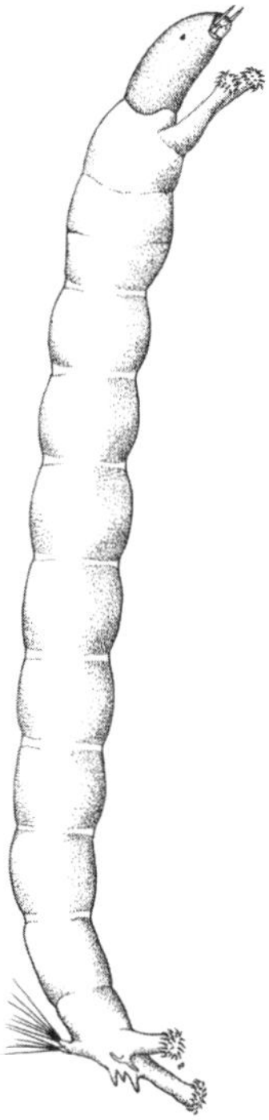

I am a freshwater-insect specialist, and my favorite group is the small but abundant group of Dipterans (true flies) called Chironomidae. In the early 1980s I was invited by the Shoals Marine Laboratory (SML) to join a group of fellow biology professors from around the country to look at marine invertebrates. I am a self-confessed geek and love finding animals and plants I've never seen before. I looked forward to this new opportunity: maybe I'd find the spider-like pyconogonids or the bioluminescent ctenophorans (comb jellies) I'd seen in textbooks. I could hardly wait to get my boots on and start collecting big and small creatures from the rocky shores.

My colleagues and I gathered at the docks in Portsmouth, New Hampshire, and boarded the sturdy research vessel used by the marine station: the forty-six-foot R/V *John M. Kingsbury*. Soon we were underway. Like many well-designed research vessels, the R/V *John M. Kingsbury* is able to withstand heavy seas, but that often means it rocks and rolls with every wave. That day the seas were a bit choppy, and we bobbed around like a tiny cork in a washing machine for what seemed like hours, although we traveled only about five miles. Eventually we reached Appledore Island and the SML, which is run by Cornell University and the University of New Hampshire. Only about a mile across, Appledore Island is the largest in a group of nine small, rocky islands and shoals. Since there are very few inhabitants, the islands offer an ideal location for conducting marine research. And of course, they are great places to look for marine invertebrates of the New England coast.

Upon our arrival at SML, we met with our instructors and guides, who gave us a tour of the facilities. It was low tide with plenty of time before dinner, so we headed to the exposed rocks and tidal pools. We spread out across the intertidal rocks where layers of organisms subjected to differing tide levels looked like horizontal stripes painted on the rocks. Within a few minutes we found barnacles and several distinct snails in the upper layers; they were used to being out of the water at high tide. Below clusters of dark blue mussels were other intertidal animals exposed only at lower tides. There were sea stars in multiple hues, pretty pink coraline algae, crabs, and feathery, brightly colored nudibranchs. Crawling between the attached algae I found the slender pyconogonids I'd been looking for. We even found ctenophorans: if it had been nighttime, they might have been glowing in the dark.

Nesting gulls were everywhere on the island. Their noisy cries punctuated the air, and some were intent on protecting their nests from human interlopers. If we ventured too close to an unseen nest, from out of nowhere, a gull would swoop down to peck at the top of our heads. We went into defensive mode. Small branches stuck on top of our hats diverted the gulls' attacks towards the twigs, not our scalps, and only a few hats got removed by dive-bombing birds.

Our initial study of rocky shore marine island inhabitants was coming to a close when I realized there might be an aquatic insect here that I had never seen before. Insects are very rare in marine habitats, perhaps because these habitats already sport significant invertebrate diversity, and there are very few niches for terrestrial organisms to claim. Adapting to a salty-water habitat from freshwater would be a major development for an aquatic insect. However, a number of Chironomidae, a group of true flies often called midges or chironomids, are one of the few insect groups that have successfully transitioned into marine habitats. At first glance, Chironomidae look like miniature mosquitoes. Perhaps you have seen them gathering around a summer evening porch light, but probably paid little attention because they were so small and not as "glitzy" as the larger insects buzzing around. Luckily, midges don't usually feed as adults so they wouldn't be the ones trying to bite you.

No matter where on the globe you go, huge numbers of midges can be found in freshwater habitats; they are important in streams, lakes, marshes, and ponds. There are more than twelve thousand known species of midges, and new species are being identified almost every week. Streams near the equator may be home to several hundred species of midges and even frigid arctic streams can have twenty species. Not only do we find numerous species where midges are present, we often encounter huge numbers of individuals. Some midge species produce two, three, or more generations per year from the same stream, pond, or lake. In some streams more than one hundred and fifty thousand midges emerge from a single square meter of stream bottom per year. Because of their great abundances, midges are major sources of food for both aquatic (fish and insects) and terrestrial (birds, bats, dragonflies, and spiders) predators. Though their sizes are tiny, by their sheer numbers they are important players in an ecosystem.

Most of midges' lives are spent as tiny larvae under water. After growing to a particular larval size, they become pupae for a couple days. Midges enter their adult stage by emerging out of pupal cases on the surface of the water. With environmental cues like light and temperature, an emerging adult in a lake floats on the surface while struggling out of its tight-fitting pupal skin (the exuvium); it's like taking off tight-fitting, wet, blue jeans. That pupal skin is left floating on the water while the adult flies off, living for a couple of weeks as it looks for a mate. In streams, midge pupae avoid being caught in the surface turbulence by rising quickly to the surface, then instantly "popping out" of their skins to fly away. As evidence of these activities, I often find pupal skins left by adult midges trapped in naturally occurring foam or scum on the edges of lakes or streams. I use those castaway skins (which we call "exuvia") to identify midge species living in freshwater situations. Looking at the tidal pools on Appledore Island, I wondered, would that same collecting technique work in a marine habitat?

Walking along the rocks with the incoming tide, I was following a ribbon of salty foam. Bending down, I scooped up the foam floating around the intertidal pools, then pulled out a plastic bag and stored several foamy handfuls that I could look at later under magnification.

Back in the lab, I put a little salty scum under a dissecting microscope. Among assorted pieces of debris were a number of Chironomid exuvia, each about five millimeters long. These pupal skins were clear and easily passed over as just more debris. However, they had empty head and thoracic portions with distinct wing pads. The empty abdomen case extended beyond, revealing a few diagnostic spines. I was able to determine that these were specimens of *Cricotopus sylvestis*. There are more than fifteen species of marine Chironomidae, but these were the first marine insects I had ever seen. Many times I'd read about marine midges but here they were, right in front of me.

It was a memorable trip to Appledore Island. To me, those rarely seen marine midges were as extraordinary as the vividly colored marine invertebrates. I suspect they are not usually found because we rarely look for them. We don't know much about their ecological roles in marine habitats, but we can be sure they are important members of marine rocky shore communities.

Chironomidae (marine)

Life history: Like other true flies, there are 4 distinct life stages: egg, larva, pupa and adult. Marine chironomids vary in durations of their life cycles just like lake species. In some arctic marine habitats, Chironomidae may take 2 years to complete their life cycles while more temperate marine species can produce several generations per year.

Larvae: Most of the chironomid life span is spent in the larval stage, during which they actively feed and grow through several instars. In marine habitats larvae seem to prefer sandy substrates with algae masses where they make a protective silken tube. They transform into pupae and wait for the proper time to emerge into the adult stage.

Pupae: Marine chironomids use tidal changes as their cue for emergence and usually pop out of their pupal skins during low tide.

Adults: Lay eggs in or near the water. Since there are a large number of marine midge species, there are a variety of approaches as adults. Some adults are virtually flightless and crawl around looking for a mate. Others stay on the surface of the water, gliding around to find a mate, while others form swarms and find partners in a large group. One characteristic of all marine midges is their short adult life spans. Many find mates and lay their eggs within a couple hours after emerging. This quick mating ability is an evolutionary advantage since low tide is only a few hours long. Females quickly lay their eggs on rocks or detritus at the water edge and then die.

Feeding: Most marine chironomids feed on algae, including diatoms, or detritus. While not well documented, some marine midges, like their freshwater relatives, are predators. Predacious chironomids hunt and consume other midge larvae or smaller invertebrates.

Habitat Indicators: Chironomids inhabit just about any aquatic habitat you can name. Among the diverse Chironomidae, there are diverse tolerances. The salt levels chironomids can survive range from fresh and slightly brackish waters to ocean waters and even extremely saline situations. Not enough is really known about marine midges to use them as bioindicators. However, it is a different story in freshwaters.

Some larvae have bright red hemoglobin that holds onto oxygen, so they can survive conditions such as lake-bottom mud where oxygen levels are very low. At the opposite extreme, other freshwater species live only in cold, pristine waters of alpine streams and lakes; for example, most members of the genus *Diamesa* will only survive in very cold, clean, oxygen-rich streams. Some species in the genus *Chironomus* or *Tanytarsus* are able to survive in highly impacted, organically polluted ponds or lakes that have only minute amounts of oxygen.

To use midges as bioindicators, it is best to look at the assemblage of species within the community. This requires examining them under microscopic magnification. If there are a large number of intolerant members like *Diamesa*, the habitat is probably pretty healthy. If the majority of the community members are ones able to survive low oxygen conditions or polluted conditions, it may indicate the presence of poorer conditions.

17 The Phantom Midges of Silver Lake

Michael C. Swift

It was mid May, and everyone was returning for the summer season at Silver Lake. I'd been sewing nets in preparation for sampling when I saw the twins, Sue and Mark, running down the dock toward me. The Franklin family had the cabin next to mine, and I always looked forward to their arrival. The twins should be about ten years old now, I mused.

"Can you take us out on the lake?" the children said together. I could see that they were ignoring their mother's calls for help unpacking the car. "We want to go fishing, but Grandpa says there aren't any fish in Silver Lake anymore."

"Hold on, hold on," I said, laughing. "First you need to get moved in, and I need to finish mending this net and organize my sampling gear. How about going out tomorrow, after lunch?"

The next day, after I loaded a few supplies and my sampling equipment, the kids climbed into my boat, and we set off for the middle of the lake. "Let's see what critters we can find with this plankton tow," I explained, handing them the gear. "We'll start with a vertical haul from the bottom of the lake to the surface. Sue, you lower the net until it's fifteen meters deep; that will put it just above the bottom. See the marks on the rope? Mark, you watch as she lowers the rope and call out the depths."

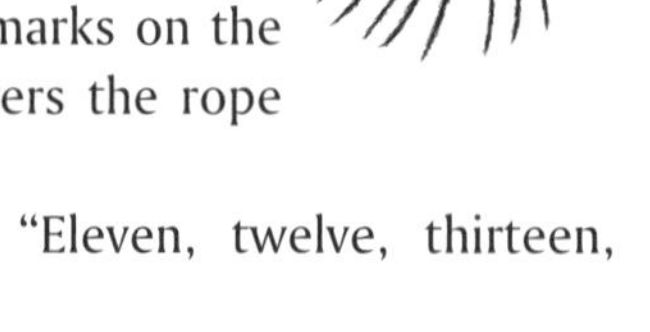

Sue began to lower the net. "Eleven, twelve, thirteen, fourteen, STOP!" called out Mark.

"Now, slowly and gently pull the net up," I instructed. "When it reaches the surface, we'll rinse it and then empty the plankton into a jar." They both performed like pros. In no time we had a jar full of plankton. "What do you see?"

"There are little critters zipping all around in the jar," Sue said as she tightened the lid.

"What are they?" asked Mark.

I held the jar up to the sky so we all could see the plankton against a light background. "It looks like copepods and cladocerans for sure. I'm also looking for rotifers, which are too small to see without a microscope, and transparent *Chaoborus* larvae. But maybe it's too early for them."

"What's a koberus?" asked Sue, quizzically.

"Ka-ob-er-us, spelled C-H-A-O-B-O-R-U-S, are fly larvae that live in the mud at the bottom of the lake," I explained. "They migrate up into the water column and come almost to the surface every night to feed. They'll catch tiny rotifers, and the bigger zooplankton like the ones you see in the jar. Once they feed, they migrate back down to the mud where they spend the daylight hours."

"Well," said Mark, "If they only come up at night, why would we catch any in the afternoon?"

"Very good question. Let's look at this plankton sample in the lab to see whether we caught any." After preserving the sample we returned to the dock and went straight to the small laboratory in my cabin. Under the microscope I could show Mark and Sue the diversity in their sample. We had collected lots of red, teardrop-shaped copepods, each with a single eye and either long or short antennae. "Those with the long antennae are *Diaptomus* and the ones with shorter antennae are *Cyclops*," I told them. "Fats stored in their bodies have red pigments that give them their color. And fats are used for making eggs. See how many of them carry clusters of eggs?"

"What are these clear, fatter critters with long arms, one big black eye, and a spine?"

"Those are *Daphnia*, Sue. Their antennae, those things that look like arms to you, are used for swimming, and they have legs to filter algae. Their outer protection is a carapace shaped like a folded taco shell. Cladocerans like *Daphnia* migrate up and down the water column, but maybe not as far as *Chaoborus*."

"Now I see something different from *Daphnia* or the copepods," said Sue. "They are long and skinny. Look, Mark."

Mark looked into the microscope. "It's really transparent, but has little black dots near both ends and strange projections at one end."

"Maybe you're looking at an early instar larva of *Chaoborus*." As soon as I looked into the microscope, I confirmed it was a young *Chaoborus* larva. "The dark spots near both ends are pairs of air sacs. They are used to maintain neutral buoyancy—this young fly larva can hang horizontally and motionless in the water, waiting to strike at the right-sized copepod or cladoceran. When the larva feels vibrations from swimming prey, it grabs the prey with its modified antennae and other mouthparts. Once the prey is securely held, the very pointed mandibles punch a hole into the victim, pushing it into *Chaoborus*'s muscular crop at the beginning of the digestive tract. Here's picture of a mandible."

Mark and Sue studied the photo closely. "It looks like a pitchfork, or a fish spear point," Mark exclaimed as he looked up at me.

I smiled and continued, "Digestion occurs in the crop, then all the liquid material moves into the rest of the digestive system. Undigestible bits are regurgitated through the mouth and the larva is ready to catch another meal."

"This larva is pretty small. How can it eat those large copepods and cladocerans?"

"This is an early instar, or growth stage—maybe number one or two of four. They'd be eating the smaller critters. Older, bigger instars eat the bigger zooplankton. We can go sampling tonight, if it's all right with your parents. We'll try to collect some older phantom midge larvae and pupae." They were eager for another trip, and we made arrangements to meet at the dock at eleven o'clock that night.

At ten thirty the twins came running down to the boat, dressed warmly and ready for action. Only when we were in the middle of the lake did I explain the night's sampling. "Okay, we're going to use a different net tonight. We expect that the *Chaoborus* larvae have migrated near the surface, so we'll drag this horizontal plankton net just below the surface. Sue, you unwind the rope while Mark lets out the net. Let about five meters of rope trail off the rear of the boat—I'll keep rowing. Mark, when you get to the five-meter mark on the rope, hold it so the net doesn't drift any further back. If I row slowly, the net will only sink a little way, and we will sample horizontally near the surface."

I rowed for several minutes. "Now slowly pull in the net, Mark, and rinse its contents into the bucket. Sue, put your flashlight under the bucket. It will light up the plankton in the bucket."

"Wow," cried Sue and Mark at the same time. "There are lots of things wriggling around, but it's hard to see just what they are."

"Exactly. Let's take a few more tows. Tomorrow, we'll look at them under the microscope." We repeated the process several more times and then headed in for the evening.

The next morning, Sue and Mark were at my cabin right after breakfast. "We're ready," they called out. "What did we catch?"

Under the microscope were the familiar copepods and cladocerans from the day before, but there were also some new animals. "Looks like the *Chaoborus* we saw yesterday, but it's really big," Mark exclaimed. "It's way longer than the copepods and cladocerans, but not much bigger around."

"Good observation," I said. "You are looking at fourth-instar larvae. They're usually about one centimeter long, but only a couple millimeters in diameter."

Sue also found new features when it was her turn at the microscope. "There's a red line down the center of the body, a tail made of tiny hairs, and what looks like bright, black eyes."

"Very good eyes, yourself," I said. "The red line is the digestive tract—it's stained by the red lipids (fats) from the copepods they have been eating. The tail hairs form a fan that hangs down below the body when it is sitting motionless in the water waiting for prey. The larvae move by bending into a loop, then straightening out rapidly. When they straighten, the tail fan pushes against the water and sends them off in a random direction. The eyes you see are compound eyes, like those in a *Daphnia*. They sense light intensity rather than visualizing prey. Remember, they strike at prey in response to water movements."

Mark saw something else: "What is this funny thing? It has big eyes, a paddle tail, and two horns sticking up out of its head."

"You found a *Chaoborus* pupa. I was sure you would, since we saw young larvae in our daytime sample. Those young instars meant there were adult *Chaoborus* around, and that pupa will be an adult soon. Fourth-instar larvae spend the winter in lakes and ponds then pupate in late spring. They'll emerge and fly in large swarms where they breed. Adults look a lot like mosquitoes, but they don't bite.

"Female *Chaoborus* lay a cluster of eggs on the water surface, and the eggs soon hatch into first-instar larvae. The tiny larvae have no compound eyes yet, but you saw that they look a lot like the older larvae. We found the young larvae in our daytime sample because the first three instars usually don't migrate vertically to go deeper in the lake."

"Why don't they migrate?"

"As *Chaoborus* larvae evolved with fish, they developed a daily migration pattern to avoid being seen, and eaten, by fishes. First-instar and second-instar larvae don't need to migrate down into the mud because they're too small to be seen by fish. There's a slim chance fish might see third-instar larvae, but fourth-instar larvae and pupae are quite visible, so they seek refuge in the lake bottom during the day.

"Your Grandpa was telling me he isn't catching nearly as many fish compared to when he was a boy your age. Without fish in the lake, big *Chaoborus* are pretty common these days. Remember the old pictures of fish and fishermen all over the walls at the Lodge? There are fewer fish because acid rain has gradually killed the fish in Silver Lake. Emissions from our driving and chemicals produced from power generation combine with rain in the sky to make the water acidic. But there's still food for them—copepods and cladocerans are still living in the lake."

"And *Chaoborus*," added Mark.

"Yes," I agreed, "*Chaoborus* can tolerate quite acidic conditions."

"If there are almost no fish, then why do plankton still migrate?" asked Sue, puzzled.

"It takes a long time for a behavioral pattern to change. It's only been a few years since fish disappeared from Silver Lake. There is talk that the Department of Fisheries and Wildlife may be treating the lake to make it less acidic. Then fish could probably live here again."

"For now, I guess we can be glad there are still zooplankton making their daily migrations in our lake," Sue sighed, looking out towards the boat launch.

"And some day, if we can restore this lake, they'll be here when you and your Grandpa can fish here again."

Chaoborus spp.

Life Cycle: Univoltine (multivoltine where temperatures are warm).

Larvae: 4 instars, length from 2 mm (1/16 inch) (instar I) to 1 cm (1/2 inch) (instar IV); highly transparent.

Pupation: Lasts about 1 week in late spring if univoltine or semivoltine; usually shorter if multivoltine.

Adults: Adult phantom midges look like small mosquitoes, but are non-biting.

Feeding: All larval instars are predators on small planktonic animals; adults feed on nectar; adults lay rafts of eggs at the edges of waterbodies after breeding in swarms.

Habitat Indicators: *Chaoborus* larvae are important constituents of lakes and ponds, where most species coexist with fish. They inhabit a unique niche because they are part of the plankton (living in the open water), and also part of the benthos (living in the mud). *Chaoborus* larvae can be an important food source for fish. As lakes acidify from the influence of acid precipitation, biodiversity is lost, leaving the habitat with only a few species, almost always including the acid-tolerant *Chaoborus.* There are several *Chaoborus* species in ponds in the Rockies and the Pacific Northwest; *C. punctipennis* is widely distributed particularly in the Midwest and East.

PART V

Dragonfly Detectives

About Dragonflies (Odonata)

Dragonflies and damselflies (Odonata) have an ancient lineage on earth, going back over two hundred fifty million years. Their colorful adult markings, large size, and conspicuous behaviors often afford them a charismatic megafauna status among aquatic invertebrates. Their name derives from the Latin *odon* meaning *toothed*; this reflects the large mouthparts of the larvae, which allows them to grab and manipulate prey. Dragonflies and damselflies are predators, both as larvae and adults. Most larvae, often called nymphs or naiads, are tied to permanent water, and the few that inhabit temporary habitats have shorter life cycles. Some larvae live on the wet margins of streams and bogs, demonstrating a quasi-terrestrial lifestyle. While some adults spend all their daylight hours cruising stream edges or perched on vegetation near water, others venture away from water for much of their adult lives.

Morphology

The simple body form of the nymphs is similar to that of adults, though the head and eyes are smaller. The distinctive characteristic of all odonate nymphs is the lower "lip" or labium. It is equipped with various hooks, spines, and teeth. Because the labium is folded under the front legs, it can be sprung forward to capture prey. Six legs, borne on the three segments of the thorax, are often robust and make these predators fairly agile in the aquatic environment. The abdomen of damselflies is more slender and longer than the form of their dragonfly cousins; damselfly nymphs are also distinguished by three external gills on the posterior ends of their abdomens known as caudal lamellae. Dragonflies have gills that are contained within the abdomen; expanding and contracting their abdomens draws water across the gills. This mechanism can also be used in propulsion—some dragonfly nymphs will rapidly expel water from their abdomens, producing a water jet that shoots them forward.

Adults have huge, multifaceted compound eyes. Each eye has approximately thirty thousand facets. Their freely movable heads further enhance the insects' ability to see prey, detect avian predators, and search for mates. Wings borne on the tough thorax are held horizontally at rest in dragonflies; in contrast, wings are held together above the body at rest in damselflies. Their wings move in a paddle-like fashion that makes odonates excellent fliers; flight for some dragonfly species has been estimated at speeds between twenty-five to thirty-five kilometers per hour. Adults breathe directly through spiracles in the thorax and abdomen.

Life History

Odonata are hemimetabolous, which means that nymphs develop gradually, without a pupal stage. The egg stage takes from eight to thirty days depending on the weather. Some species undergo diapause during the winter. After hatching, it takes from one to three years (for some rare species, five or more years) for them to undergo ten to fifteen molts, gradually developing wings in the final instars. Their hinged labium makes them highly successful predators of other invertebrates and young fish. The various families in the order have particular predatory strategies: burrowers that lay and wait for their prey, sprawlers that more actively hunt in sediments or organic debris, and climbers that lurk in vegetation to stalk their prey. The nymphs are non-discriminating predators, and they will eat any other invertebrates in the water. At times, these nymphs may be among the few remaining taxa in a habitat, and as a result they prey upon themselves.

When nymphs crawl out of the water to emerge, they cling to aquatic plants while their exoskeletons split and blood pressure fills out their wings. This is a vulnerable time in their life history, and emergence often occurs at night to avoid predators. At first their color is pale, but eventually mature body colors develop, with the male often appearing much more conspicuous than the female. Adult behaviors, as with the nymphs, vary greatly, from those that cruise constantly in search of prey, to some that perch on vegetation by hanging vertically, to others that perch horizontally and snatch flying prey like flycatchers. Adults generally live from two to four weeks, culminating in mating activities that might include continuous tandem attachment of the male and female, even during egg deposition into the water or on aquatic plants. This behavior is a form of mate guarding, by which males prevent access to the female by males of the same species.

Bioindicators

Odonates do not require complex habitats and can adjust to most any condition. The nymphs of dragonflies and damselflies are not very reliable indicators of water quality because they are not usually directly affected by chemical contaminants. In general, they are moderately tolerant and range in distribution from good conditions to all but the most severely impaired systems. While most species are generalists, those that have very specific habitat needs in their larval stage can be indicators for the quality associated with particular habitats.

References

Corbet, P. S. 1999. *Dragonflies: Behaviour and Ecology of the Odonata.* Ithaca, NY: Cornell University Press.

Gordon, S., and C. Kerst. 2005. *Dragonflies and Damselflies of the Willamette Valley, Oregon.* Eugene, OR: CraneDance Publications.

Kerst, C., and S. Gordon. 2011. *Dragonflies and Damselflies of Oregon: A Field Guide.* Corvallis: Oregon State University Press.

Hudson, J., and R. H. Armstrong. 2005. *Dragonflies of Alaska.* Anchorage, AK: Todd Communications.

Tennessen, K. J. 2008. "Odonata." In *An Introduction to the Aquatic Insects of North America,* edited by R. W. Merritt, K. W. Cummins, and M. B. Berg, 237–43. Dubuque, IA: Kendall Hunt.

18 Hanging from a Leaf

Rob Cannings

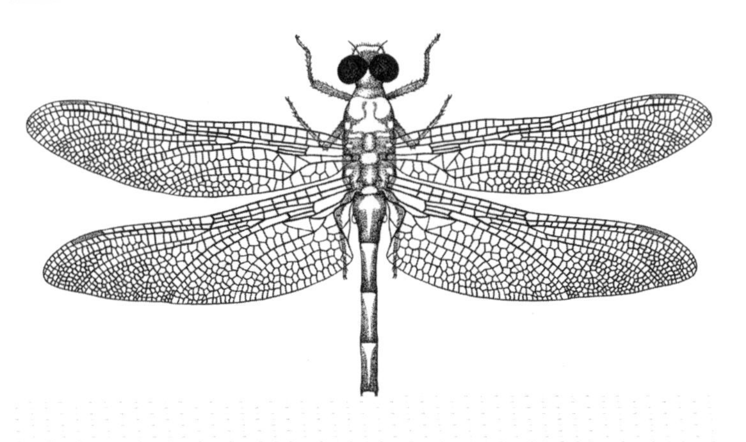

I grew up along the Okanagan River in Penticton, British Columbia, and, when I return to my hometown, I walk its dikes, watching the mergansers on the river and listening to the catbirds and orioles in the dogwoods and cottonwood trees. I'm an entomologist, though, with a special interest in dragonflies, and I love to keep track of these bold, beautiful insects along the river. The rarest of the rare here is the olive clubtail, *Stylurus olivaceus*. In the late 1920s, decades before the river was straightened, dredged, and dammed to control flooding, my father used to paddle his homemade kayak on the river at Penticton. When I was a kid, he'd tell me of his adventures there. "Birches hung over the gravelly riffles and willows lined the sandbars along the slower flowing, meandering reaches," he'd say. "We'd explore all day in the woods along the shore, in the cattail marshes and wet meadows that lined the old oxbows." I'd listen raptly to his boyhood stories of a landscape now almost completely gone.

The Okanagan River empties out of the eighty-mile-long Okanagan Lake at Penticton in southern British Columbia and winds south to join the Columbia River in Washington State. Now, much of the land along the river is rich farmland or urban development, but the river used to flow through riparian woods in sage and antelope-brush grasslands. In the Canadian part of its journey, most of the river has been channeled since the 1950s; these days it runs between dikes used as walking and bicycle paths.

The first record of olive clubtails on the Okanagan is found in the writings of Frank Whitehouse, an avid dragonfly collector, who found the insect scarce in July 1938. The males he saw "pursued a zigzag course in the middle of the river—carefully avoiding my boat." He continued in frustration:

> Then I came equipped with a bathing suit and tried standing mid-stream on silt bars. No *S. olivaceus* would then appear! Once, in the boat, I was within six feet of a fine male. They do not appear to take up any limited reach; but approach, go by, and continue going!

If Frank had searched in late August or early September, he probably would have see many more; July is usually too early for clubtails to appear in the Okanagan Valley.

Today, seeing an olive clubtail along the Okanagan River anywhere in Canada is cause for celebration because they are now uncommon. The channel bottom has been lined with rocks and boulders, making the sand and silt that clubtail larvae need for burrowing scarce and patchy. For evidence of larvae, I often look for an exuvia, the cast skin of the last larval stage, which remains clinging to shoreline debris after an adult insect has emerged and flown away. I've never found exuviae on this river. And the trees and shrubs beside the water, where the adult dragonflies love to perch, are mostly gone, now replaced by grasses and weeds.

In all my Okanagan River dike walking, I've found only three olive clubtail adults. The first one I'd ever seen was camouflaged like a grey-green and black twig, perched flat on the dusty trail at Osoyoos Lake, almost at the U.S.–Canada border. What excitement! On another walk much later, I saw one flying over the river, back and forth. After a while it flew towards a lone tree on the dike. "Aha!" I thought, "It's landed there!" *Stylurus* dragonflies are referred to as "hanging clubtails" because they normally perch on leaves of trees or shrubs beside the river, bending the leaves until they are hanging almost vertically. Sneaking up quietly, I searched the lower branches for five minutes before I saw the dragonfly hanging from a leaf, just as it was supposed to do! My third sighting was of a male flying up the river, fast and straight as an arrow, not stopping for anyone.

Most of the olive clubtails I've studied belong to a completely separate population. They live at the most northerly place the species is known: the Thompson River near Kamloops, about eighty air miles northwest of Penticton. For about thirty-five miles east of the city, this big river flows through sagebrush, farms, and riverside suburbs, its water warmed to 21–22°C in summer after its stay in the huge Shuswap Lake to the east. At Kamloops it's joined by the North Thompson River, with its colder waters that drain mountain snowfields. Here the water is about 18°C at the time of dragonfly emergence, and I suspect the water is too cold for olive clubtails at that time of year. I've never seen this species in the North Thompson, even though other conditions such as sediment and stream flow would seem perfect for the clubtail. Downstream from Kamloops, the Thompson River flows faster and the riverbed is filled with boulders which, as I knew from the Okanagan River, is a habitat disliked by these dragonflies.

It's ironic that one of the best places anywhere to find the olive clubtail—a lover of warm climates—is as far north as they can live. But here much of the habitat is still good with a sandy, silty riverbed, stable banks clothed with emergent water plants such as rushes and horsetails, riverside willows, and introduced Russian olive trees hanging over the water. Certainly, there are stretches where livestock have trampled the shore, where irrigation water has eroded the banks, and where rocky fill has been dumped to support the railway line and subdivisions full of houses. But there is still enough good shoreline to support a decent population of dragonflies. In some places upstream of Kamloops, I picked up one exuvium for every yard of sandy riverbank I walked; they lay among the detritus of high water in the patches of horsetails and rushes, coated in a thin layer of silt, reminders of the larva's burrowing life.

Luckily, the habitat does not need to be pristine: the clubtail can tolerate some disturbance and habitat damage. I found exuviae in the imprints of cow hooves and saw some adults emerging at a busy boat launch. But I imagine carp and other introduced, bottom-feeding fish can hurt the population by disturbing the bottom mud or by actually eating the larvae. Additionally, motorboats speeding by stir up the bottom silt and erode the sandy banks with their wakes.

These disturbances notwithstanding, along this reach of the Thompson, in mid August, the adult dragonflies emerge, pale and vulnerable, easy prey for blackbirds and kingbirds. I suppose the clubtails fly back in the grasslands to hunt and mature for several days, later returning to the river to mate and lay eggs. In late August and September, now and again, I've watched a male patrol over the current, chase down a female, and then link with its partner in the loop position so unique to mating Odonata. Flying back to the shore, the pair disappears into the trees. A couple of times I've been lucky and have found them hanging from a low twig. Later, the female flies fast out over the river, dipping her abdomen into the water and washing off the eggs.

Because these are small and fragmented populations, and the few stretches of rivers where they live are vulnerable, the olive clubtail is considered a threatened species in British Columbia. Its future is of less concern in the western United States, where it is more widespread. In some places, such as in the Lower Columbia River, it is common. Downriver from Portland it even lives in waters affected by the tides where it seems to tolerate some salinity. The olive clubtail is characteristic of the big rivers of the dry West; although in many places it's been hit hard by the upheavals humans have brought to its home and is now rare, it remains common in some localities. To me, as I consider the populations I have studied in British Columbia, the olive clubtail remains a symbol of perseverance in a rapidly changing landscape.

Stylurus olivaceus

Life Cycle: Probably a 2-year life cycle, as is common in other temperate-zone clubtail dragonflies studied. About 12–15 molts from egg to adult.

Larvae: Elongate with the abdomen only a little wider than the head; the abdomen gradually tapers to the pointed tip. Fully grown, they reach almost 1.5 inches long. They burrow in sandy/silty river bottoms.

Pupae: No pupal stage. Larvae crawl from water when mature; adult emerges from larval skin (exuvia), a process that takes about an hour.

Adults: Medium-sized dragonfly, about 2.25 inches long with a wingspan of about 3 inches. Pale areas dull green marked with brown/black; wide shoulder stripes; abdomen mostly black with pale, spear-shaped marks on top; eyes blue in life. This is the only clubtail dragonfly in the region with minimal striping on sides of thorax. Most emerge beginning in June in the southern part of its range (California), July or early August in north (Washington and British Columbia). Adults live 6–8 weeks and fly through September or later.

Feeding: Larvae are predators of insect larvae and other invertebrates in river sediments; adults catch and eat flying insects.

Habitat Indicators: Prevalent in warm, sandy/silty-bottomed, medium-sized to large rivers, typically with sandy banks, and warm lowland habitats, mostly east of coastal mountains. Rivers flow through grassland and sagebrush habitats or through woodland. These habitats now are often modified by humans, but the species can tolerate some habitat disturbance. Needs riverside shrubs and trees for perching. Requires relatively stable riverbanks.

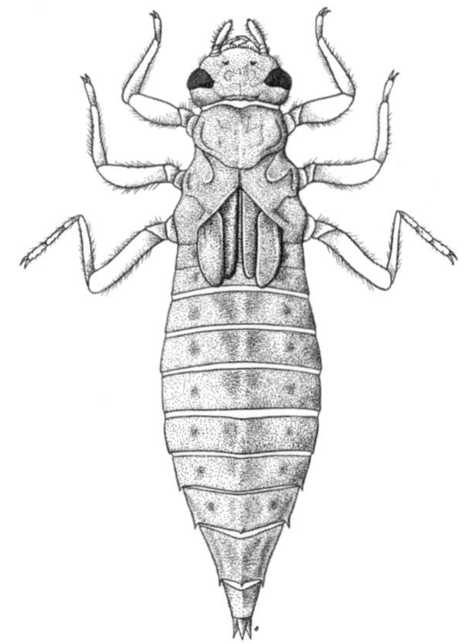

19 Tracking the River Cruiser

Christopher Beatty

It was coming straight at me. I stood there, in the middle of the field, bracing for it, getting ready. "Keep breathing," I thought to myself, "swing quickly, follow through . . . this might be the only chance you get." I could see the sunlight glinting in its eyes as it bore down on me. Two meters . . . one meter . . . it was there . . . SWOOSH.

I missed. The dragonfly swerved, bounced off the rim of my net, regained flight, and dashed away toward the edge of the field. I took off after it—I knew I had no chance of running it down, but it might change direction when it reached the edge of the clearing, and maybe I could get out in front if it again. What I did not expect was that, at the edge of the field, it would turn around and come straight back at me once more. This was my second chance . . . ready . . . SWING.

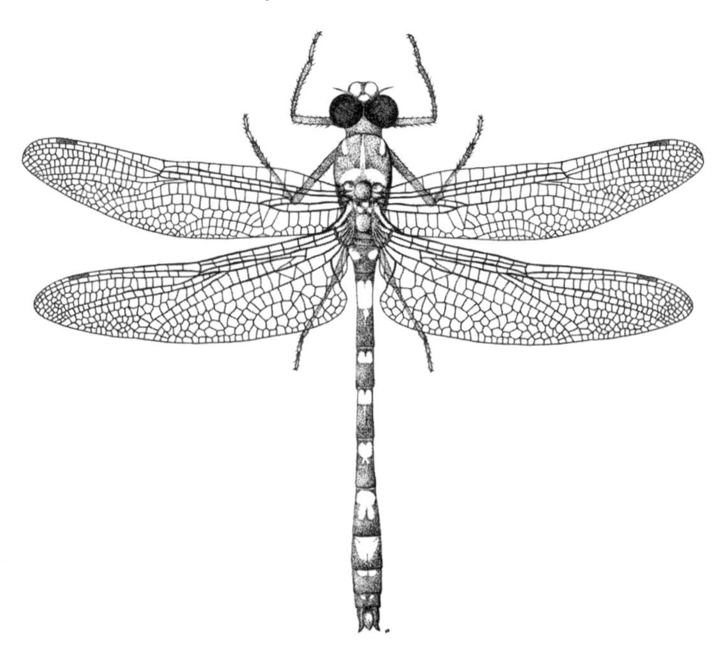

The hiss of the net moving through the air finished with a quick twist of the handle to close the net bag. A fraction of a second passed, during which I heard only the rustle of the breeze in the grass around me, and then from inside the bag came the telltale roar of fast-moving wings. I had captured my first *Macromia magnifica*, the western river cruiser.

Steve Valley, Odonate expert extraordinaire and my frequent companion during these collection expeditions, came running up behind me, slightly out of breath from dashing across the open meadow. He crashed through the tall grass, the bag of his own net trailing behind. "Did you get it?" he called out as he approached.

"Yes!" I replied as I slowly reached into the net to retrieve the dragonfly.

Steve came to a halt in front of me. "Is it a *Macromia*?" he asked excitedly.

Reaching into the bag, I removed a large dragonfly: dark body with pale yellow stripes, a yellow leading edge to the wings, and violet-blue eyes. "Yes," I replied again, as I held it up.

Then came what I have since referred to as my "Jedi moment" for dragonfly work: Steve looked at me thoughtfully and said, "Well, Chris, you can bag a *Macromia* on your own now: I have taught you all I can."

This was hardly an accurate statement—Steve's extensive knowledge of dragonflies had been developed through a lifetime of fieldwork and reading, and if I could ever hope for something even approaching his level of expertise, I would need to spend years at work, as well. This was the first summer of fieldwork for my master's thesis, studying the distribution of dragonflies and damselflies in riverine wetlands in the Willamette Valley of western Oregon. Steve had taken me under his wing, visiting study sites with me, discussing species identification and dragonfly behavior and ecology. I had a lot to learn about dragonflies and what they could tell me about wetland ecology. That first summer my learning curve was steep.

Truth be told, that day had begun as a bit of a break from my field research—an afternoon spent playing "hooky" from studying dragonflies . . . to go catch dragonflies. This open meadow along the Coast Fork of the Willamette River in western Oregon was high and dry, standing at the foot of Mt. Pisgah, just a few meters from the river channel: not one of the floodplain wetlands that I was studying. My field sites had a mix of different dragonfly species—the black and white skimmers of *Libellula* and *Plathemis*, the bright red *Sympetrum* meadowhawks, the greens and blues of the western pondhawk (*Erythemis collocata*) and the Blue Dasher (*Pachydiplax longipennis*). These were species that laid their eggs in the quiet waters of wetlands and ponds. Their larvae could live in the muddy bottoms of these habitats, foraging among the bases of sedge or hunkered down in the mucky bottom, waiting to prey upon crustaceans and the larvae of other small invertebrates.

Macromia was a different beast altogether. Larvae of many species of dragonflies and damselflies live exclusively in moving water, among the sands, cobbles, and large woody debris in streams and small rivers. *Macromia* likes moderate- to fast-moving water, such as bigger rivers like the Coast Fork of the Willamette. Its larvae look very different from still-water species, flat and wide with long sprawling legs—almost spider-like—they grip onto the substrate in fast current and move around in pursuit of prey, staying low and avoiding the faster velocity of the water all around. To find *Macromia* here told me something about this river and about its health: as an invertebrate top predator, dragonflies reflect overall conditions in their habitat, and their absence can indicate problems further down the food web.

The adults of many dragonfly species move far and wide, cruising across the landscape to find new habitats, while others stick close to home, laying their eggs in the same streams and ponds from which they emerged. Looking into a local pond or river and finding which species live there can tell us a great deal about the habitat: water chemistry, cycles of drying, structural components, all of these aspects of habitat are reflected by the organisms that live there. Thus, the *Macromia* I caught that day told me something about the fast-flowing, clear waters of Coast Fork nearby.

Since then, my research has taken me outside the Willamette Valley to other places in North America, as well as to more distant locations like Fiji and Peru. At every place I visit, I encounter a species that is new to me. I've discovered that, like *Macromia*, they all have something to say about the habitats in which they live. Thus, my quest to become a dragonfly "Jedi" continues.

Macromia magnifica

Life Cycle: Univoltine. Larvae live for 8–10 months, feeding throughout the year. Adults emerge and fly for a few weeks, between May and September.

Larvae: Large, flat, and somewhat round, with long legs (10–20 mm). These larvae are predators, moving over the sand and rock surfaces of the stream bottom to find prey.

Adults: Short-lived (3–6 weeks), foraging in the afternoon until late twilight over river pools and open meadows.

Feeding: Predators in larval and adult forms, feeding on larvae and adults of smaller insects.

Habitat Indicators: Moderate to large streams and rivers, with medium to rapid flow rates, are the best places to search for *M. magnifica*. Muddy or sandy streambeds are the preferred larval habitat, though some can also be found in rocky streams. *M. magnifica* can also be found in canals and lakes, especially in the northernmost reaches of its range. Adults fly in open areas adjoining streams and rivers, and can be seen flying along forest roads. The distribution of *M. magnifica* ranges from northern California to Nevada, eastern Oregon and the Willamette Valley in western Oregon, eastern Washington, and southern British Columbia.

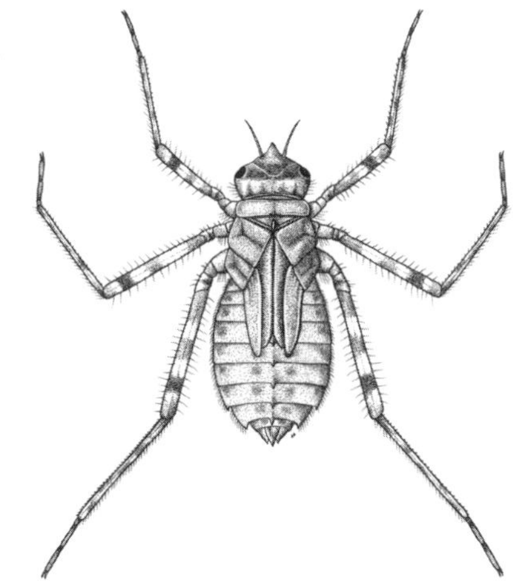

PART VI

Bugs and Beetles on Their Best Behavior

About Beetles (Coleoptera)

Beetle species (about three hundred and fifty thousand) make up the largest order of life on earth. Aristotle gave them their name from the Greek words *coleo* meaning shield, and *ptera* meaning wing. About five thousand species are aquatic and occupy just about any aquatic habitat imaginable—from running fresh waters to salty marine or estuarine systems—while others are semi-aquatic living along shorelines, deep in stream subsurfaces, in reeds of rooted plants, or in rotting logs. Unlike other aquatic insect orders, beetles are not the dominant organisms in these habitats. They generally live on top of substrates, and the ones that are good swimmers have to return to the surface frequently to replenish their air supplies. They are an ancient group and have invaded aquatic habitats many times over evolutionary history. Beetles diversified greatly during the Jurassic (210 to 145 million years ago), when the whirligig beetles (Gyrinidae), scavenger water beetles (Hydrophilidae) and diving beetles (Dytiscidae) appeared. In short, there has been plenty of time for this order of insects to diversify.

Morphology

Though their morphologies can vary greatly, many larval beetles are elongate and slightly flattened, a shape that helps them hide in crevices or adhere to rocks. They have distinct heads protected by a heavy cuticle, with strong mouthparts that scrape algae from rocks or capture prey. All aquatic beetles (except weevils) have three pairs of claw-bearing legs on the thorax. Many have gill-like appendages along the abdomen or a pair of hardened appendages at the terminal end.

Adult beetles are generally long-lived (often an entire season), an unusual trait among aquatic insects. They have a pair of hardened forewings, called elytra, and most have a second pair of wings for flight folded under the elytra. Though some beetle families live their entire lives in the water, most adults take flight for dispersal at least once in their lifetimes. Many adult beetles deploy chemical defenses that are distasteful to potential predators.

Beetles have developed several ways to breathe underwater. Some hold air reserves under their wings; some breathe through their cuticle, with or without gills; some possess unwettable hairs that hold air against the body's undersurface (called a plastron); while still others pierce plant tissues for air.

Life History

Most aquatic beetles undergo three to eight larval instars. As with the diversity in their morphologies and lifestyles, the time spent as larvae varies between only a few weeks to the majority of the year. For example, water pennies (Psephenidae) are larvae all year, but adults are found for only a brief time in the summer; in contrast, riffle beetles (Elmidae) are collected as larvae and adults almost any time. For many beetles, larval development takes six to eight months, followed by a short pupation in terrestrial habitats such as under stones or logs. Often they emerge as adults in two to three weeks.

Bioindicators

Beetles can be indicative of particular habitat conditions associated with how they breathe. Those that acquire oxygen through a plastron, such as riffle beetles (Elmidae), require highly oxygenated, usually swiftly moving waters. Those that return to the surface regularly to resupply a bubble held under their wings (like big, predatory dytiscids), must stay in waters near the edge, calm enough for resurfacing. Even more closely tied to stream edges are minute beetles that live in the small spaces between well-aerated sand particles. Other semi-aquatic beetles are associated with muddy substrates.

References

White, D. S., and R. E. Roughley. 2008 "Aquatic Coleoptera." In *An Introduction to the Aquatic Insects of North America*, edited by R. W. Merritt, K. W. Cummins, and M. B. Berg, 571–75. Dubuque, IA: Kendall Hunt.

Liebherr, J. K., and J. V. McHugh. 2003. "Coleoptera," In *Encyclopedia of Insects*, 209–230. V. H. Resh and R. T. Cardé, eds. Orlando, FL: Academic Press.

About True Bugs (Heteroptera)

Aquatic true bugs belong to the suborder Heteroptera, order Hemiptera. There is great diversity among these insects, but they are all equipped with a slender sucking tube or "beak" used to extract fluids from animals or plants. The word "bug" derives from the Middle English word "bugge" meaning "spirit" or "ghost," and was originally associated with the bed bugs that disappeared in the morning after biting their human victims during the night. Few aquatic true bugs bite humans, but many are predators on other invertebrates. Some are semi-aquatic, such as water skippers (Gerridae) and riffle bugs (Veliidae), living on the surface or near water, but never in it. A larger number are truly aquatic, such as water boatmen (Corixidae) and giant water bugs (Belostomatidae); these live under water. Aquatic true bugs range in size from tiny, one-millimeter velvet water bugs (Hebridae) to the biggest of all aquatic insects: the giant water bugs that can grow to 112 millimeters long. There are about thirty-eight hundred species of aquatic or semi-aquatic Heteroptera in the world, and only 412 species known in North America. Though true bugs in general are not well understood, some of the highly visible aquatic bugs have been studied in greater detail.

Morphology

Aquatic true bugs are dark on their top surfaces and light on the surfaces facing into the water. The elongate beak or rostrum, which true bugs use to suck plant or animal tissues, is positioned at the front of the head. The three segments of the thorax are fused and difficult to see. These bugs have a distinctive triangular structure called a scutellum (a "little shield") that is a modification of the middle of the thorax (the mesonotum). Three pairs of legs are often highly specialized (e.g., for grasping prey, skimming on the water surface, or paddling underwater).

Most Heteroptera have wings, and they disperse primarily to find mates. The forewings are half slightly hardened and opaque, half membranous. However some groups, such as water skippers, adapt to temporary habitats with populations of two types of individuals: those that save energy by not developing wings, and winged individuals that can fly to new habitats. Other species, such as the giant water bugs, have no winged individuals.

Like beetles, aquatic true bugs that live underwater have several adaptations for breathing. Giant water bugs and water scorpions (Nepidae) have "airstraps" or siphons that they can point up above the water surface and draw air into a bubble below their wings. Some bugs rise to the surface frequently in order to capture a bubble they carry around for oxygen. Others have special water-repelling hairs that hold water away from their body; a permanent film, primarily nitrogen gas, is formed that allows constant diffusion of oxygen for the bug. This permanent bubble is called a plastron, and helps the insect stay underwater a long time.

True bugs have scent glands and are attracted by sex pheromones for mating. Some, like water skippers, are not eaten by fish and probably have defensive secretions that make them unpalatable. Water skippers and riffle bugs move swiftly across the water surface using secretions from their mouths that lower the surface tension.

Life History

True bugs exhibit gradual development, that is, they are hemimetabolous with three life stages: egg, nymph, and adult. This suborder is the most diverse among all hemimetabolous insects. Most aquatic true bugs are nymphs during warmer months and then spend the winter as adults who lay eggs in the spring. There are generally five instars during which nymphs shed their exoskeletons to grow; wing pads develop in later instars. A useful characteristic that distinguishes immature true bugs from adults is that the nymphs have one-segmented tarsi (the last segments of their legs). Younger instars usually share the same food sources as adults, though prey they select may be smaller.

Bioindicators

Most true bugs require water conditions that allow them to swim to the surface to get oxygen, and they are often found on the edges of streams, lakes, or ponds. Because they can carry their air source with them (see notes about their adaptations above), they can be fairly tolerant of lower oxygen levels.

References

Polhemus, J. T. 2008. "Aquatic and Semiaquatic Hemiptera." In *An Introduction to the Aquatic Insects of North America*, edited by R. W. Merritt, K. W. Cummins, and M. B. Berg, 385–423. Dubuque, IA: Kendall Hunt.

Resh, V. H., D. B. Buchwalter, G. A. Lamberti, and C. H. Eriksen. 2008. "Aquatic Insect Respiration." In *An Introduction to the Aquatic Insects of North America*, edited by R. W. Merritt, K. W. Cummins, and M. B. Berg, 39–54. Dubuque, IA: Kendall Hunt.

Schaefer, C. W. 2003. "Prosorrhyncha," In *Encyclopedia of Insects*, edited by Vincent H. Resh & Ring T. Cardé, 947–64. Orlando, FL: Academic Press.

20 In Defense of Whirligig Beetles

Fred Benfield

When I began graduate study, I was interested in developmental biology and planned to pursue a PhD in botany at Virginia Tech (then known as Virginia Polytechnic Institute). I had been teaching at a junior college in Georgia for a couple of years and needed a job before starting my program in the fall. As luck would have it, I was offered a summer job by Dr. Stuart Neff at VPI, who was working on a National Science Foundation grant with Dr. Jim Wheeler, a natural products chemist from Howard University. Neff's two students, George Simmons and Bruce Wallace, were collecting and culturing several species of pest insects (tent caterpillars, face flies, and corn earworms) from which they hoped to extract and identify sex attractants or other

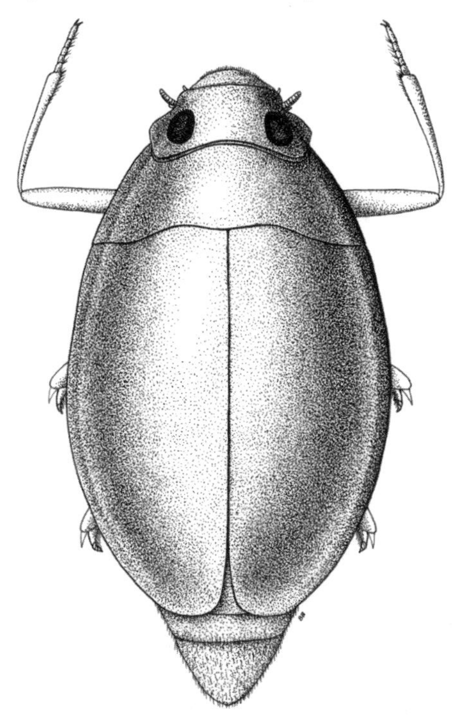

pheromones. Because Simmons was busy working on his PhD research in limnology and Wallace was trying to complete his dissertation on wetland flies by summer's end, I took over the collection and culturing duties of the project.

Toward the end of summer, we began working with defensive secretions of a group of snail-eating beetles of the family Carabidae (ground beetles). These beetles have mandibles adapted to feeding on snails, and they normally hunt at night. So we wandered around the mountains of Virginia, West Virginia, and North Carolina in the middle of the night looking for these unusual beasts. When caught, the beetles could be induced to squirt their defense secretions into a vial containing an organic solvent. Wheeler had a mobile laboratory outfitted with analytical instruments where his students began preliminary chemical analyses of the secretions. But after a time we found that field collection of the secretions failed to yield sufficient material for the definitive analyses. So I became a "beetle keeper" in order to collect the secretions from "kept" beetles over time.

I began hunting snails to feed the beetle colony, which I kept going while I "milked" the beetles for their defensive chemicals. When the secretions were collected, I packed them in dry ice and mailed them to Wheeler for analyses. Throughout the ensuing fall, I continued working with the colony, restarting it the following spring, summer, and fall. After spending my first summer at VPI hanging out with Simmons and Wallace, occasionally helping them with their research and, in the following year, going into the field to collect with Neff, I gladly chucked the idea of studying developmental biology and decided that aquatic biology would be my career path.

Stuart Neff was an entomologist specializing in aquatic groups; by the time I joined his lab, his interests had turned towards

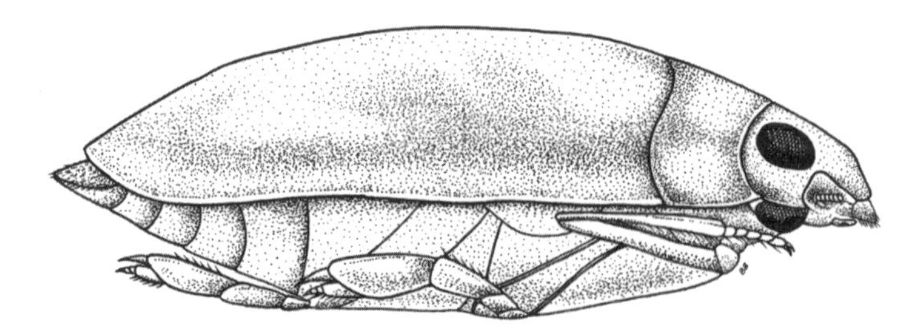

a group of aquatic beetles that had long been connected to a familiar odor: the scent of apples. Among aquatic beetles, the Gyrinidae (apple-bug or whirligig beetles) are probably the most visible because they swim on the surface of lakes and streams. During the previous winter, I thought I had settled on midges for my doctoral dissertation; however, as I started learning more about gyrinids, my fascination grew and I changed directions, working full time on gyrinid secretions and attendant behaviors.

There are four genera in the family Gyrinidae, and the genera *Dineutus* (forty-one species) and *Gyrinus* (fourteen) make up the bulk of the North American species. Individuals of *Dineutus* emit a sweet-smelling odor reminiscent of apples when disturbed. *Gyrinus*, on the other hand, emits a pungent sulfurous odor when disturbed. Because gyrinids live on the water surface of ponds, lakes, and streams, they are highly exposed to potential predators from above and below. Yet they are frequently found in groups ranging from a few individuals to massive, single-species colonies numbering in the thousands.

Because we were interested in defensive secretions, we tried to identify the chemical nature and function of the secretions *Dineutus discolor* produced. We knew the secretions were produced in a pair of pygidial glands that open in the membrane of the last abdominal segment. At first, we tried a chemist's approach—-to extract the secretion from large numbers of individuals in a solvent. Later, we changed the strategy to "milking" the secretion from individual beetles—a biologist's approach that proved to be a more efficient method of collection. The secretion was identified as a mixture of C_{14} compounds including a novel one, which the chemist named gyrinidone.

As the chemist worked toward identifying the secretion, I tried to find out how the secretion was used by the beetles. I began by conducting the simplest and most obvious test: to see if predators would eat the beetles. I put a bluegill sunfish in an aquarium and dropped in a live beetle. The fish snapped up the beetle in seconds. Boy, was I disappointed! I dropped in a second beetle—same result. However, in the third trial, the fish held the beetle in its mouth for a moment and then spit it out. I continued to drop in fresh beetles to this fish, and it rejected them all. I let the fish rest for a few days, feeding it trout chow, and then offered more beetles—all were rejected or ignored.

Next I played a dirty trick on the fish: I coated trout chow pellets with the secretion and dropped them in the tank. The coated pellet, a second, and a third were quickly rejected. Finally, the fish ignored the pellets. I repeated this series of experiments with several fresh bluegills and also with several rainbow trout—all resulted in rejection of the beetles and food coated with beetle secretion. At this point, I was pretty convinced that the secretion was effective in defense against fish, but I tried one more experiment with an aquatic predator: a red-spotted newt. I fed small bits of liver to several newts for a few days, then coated the liver with the secretion and offered it to the newts. The newts violently rejected the coated liver. Then I was convinced that some part of the secretion had a defensive function. However, that was not the whole story.

When beetles swimming on the water surface are approached, they disperse rapidly in a "fright" or "alarm" reaction. The "apple odor" can frequently be detected as the beetles disperse; the logical conclusion is that the beetles release the secretion when threatened. However, animals with defensive secretions tend to release them as a "last resort" after other defensive mechanisms fail. Possibly the secretions released by *D. discolor* under attack might trigger alarm in the group and also cause the members to disperse.

Before getting into the possibility of an "alarm" function for the secretion, it is useful to consider the physical and behavioral attributes that beetles have evolved to help them avoid predation. First, each compound eye is divided into a submerged and an aerial part, suggesting that the insect can see both above and below the water surface at the same time. Second, the antennae are specially constructed so that one part floats on the water surface while a second part sticks up in the air. The two parts are articulated so the beetles can detect waves generated by struggling potential prey (e.g., a moth or fly caught up in the surface film) or approaching predators. These beetles can also avoid bumping into things by sensing the return of waves they generate by swimming, much as bats use sonar to "echolocate." The upright portion of the antenna is likely used to sense chemicals in the environment. As a behavioral defensive ploy, when the beetles are resting quietly upon the water's surface, there are always a few individuals around the perimeter of the group, "standing watch." If the group is approached, the "scouts" swim into the group and alarm spreads like a wave; one can hear a "click-click-click" sound as the beetles bump into each other. Many individuals dive, taking down a bubble of air with them, and remain submerged for some time. Gradually, the group reassembles at virtually the same spot it occupied prior to the disturbance.

Could the beetles be using one of the techniques to produce an alarm reaction? In one of my experiments, a drop of the secretion placed on the water surface spread rapidly, then seemed to disappear (volatilize). This characteristic seemed to jibe with what I knew about other alarm substances; for example, in ant secretions, in which the chemical elicits an alarm response, then disappears. I tested the response of the beetles to introduced secretion by constructing a flow-through test chamber consisting of two cells (clear quart cottage cheese containers) coupled by two-inch-diameter hard plastic tubes connected at 90° in the middle. In that configuration water flowed from Cell 1, through the lower half of the connecting tube, into Cell 2, and out of the system through an exit tube. The water surface in the two cells was retained at the same level, but the beetles were confined to each cell by a mesh barrier at the ends of the tubes. The beetles could not see beetles in the other cell, but I could observe them through a system of mirrors. A trial began by placing an equal number of beetles in each cell and waiting for them to become quiet. Then I disturbed Cell 1 with a hand wave and observed the reaction of Cell 2. Over many trials with many groups of beetles, individuals in Cell 2 always exhibited alarm. They were detecting a chemical signal from Cell 1.

My last experiment involved putting beetles only in Cell 2, and then placing droplets of secretion on the water surface in Cell 1. The secretion dispersed rapidly over the surface and was carried into Cell 2, whereupon the beetles demonstrated the alarm reaction. After the beetles settled down, I gave them the hand wave and they also demonstrated the same alarm reaction. As before, the result was the same with many groups of beetles and many trials. I subsequently tried other water-insoluble chemicals to see if the alarm reaction could be elicited and all failed. My conclusion was that the secretion manufactured by the pygidial glands of *D. discolor* functioned in both an alarm and defense capacity.

Gyrinids have adapted to living in large groups in very exposed habitats, perhaps to impart group "security in numbers." However, they have also evolved unique physical, behavioral, and chemical defensive mechanisms that protect individual beetles. Predators learn to avoid distasteful prey but new predators recruited into populations with each reproductive cycle have to learn the same lessons. For protection against those naïve predators, these beetles can sense approach of potential danger, and communicate their alarm effectively to others in their group.

Dineutus discolor

Life Cycle: Aquatic throughout most of their life span. 1 generation per year; eggs are attached to vegetation or other objects in early spring. Larvae are present through the summer; larvae then leave water and pupate on land. Adults overwinter in leafy or woody debris on the bottom of streams.

Adults: Live 1–2 years. They move by backswimming on water surfaces of streams, lakes, ponds, and wetlands. Often form massive aggregations.

Feeding: Adults and larvae are predators.

Habitat Indicators: Because adult gyrinid beetles are restricted to the surfaces of water bodies, they require relative flat, non-turbulent surface conditions such as the edges of streams and ponds.

21 The Bugs Famously Known as Ferocious Water Bugs, Giant Water Bugs, and Toe Biters

David A. Lytle

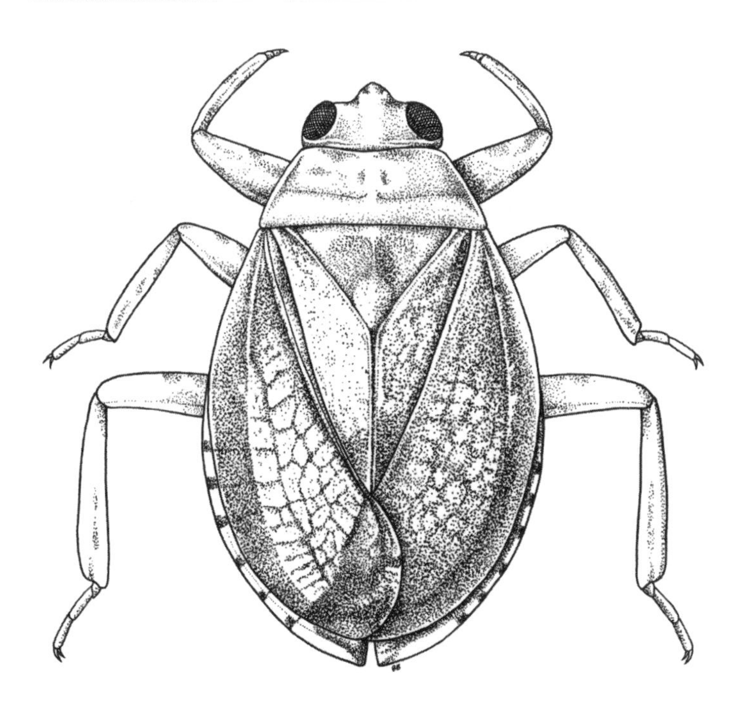

Who would expect to find the largest of all aquatic insects in the smallest of all streams? When desert streams in Arizona and Sonora, Mexico, dry down to knee-deep puddles of water in the summertime, these isolated, condensed pools are filled with voracious predators: dragonflies, aquatic beetles, frogs, and fish. But the undisputed king of this aquatic jungle is the giant water bug, *Abedus herberti*, which is particularly well suited to surviving in these unpredictable circumstances. *A. herberti* belongs to the giant water bug family, Belostomatidae, which includes the largest of all aquatic insects, with some species exceeding ten centimeters in length.

During the summer monsoon season in the Sonoran desert, colossal thunderstorms can blast a stream with intense rainfall, sending a wall of water careening down a canyon. These flash floods uproot trees, dislodge boulders, and, not surprisingly, destroy most of the aquatic insects unlucky enough to be caught in the flood's path. Years ago, I found myself standing in a stream pool during one of these downpours. Two undergraduate students were helping me collect caddisflies for a lab experiment, but the rainfall was so heavy I could barely see the stream at all. I knew a flood was imminent and feared that the stream would soon be scoured clean of insects. We were soaking wet and freezing cold: we had to work fast. Furiously, the students and I collected caddisflies, and after about fifteen minutes, we were ready to evacuate. We scooped up our gear, secured our collections, and started to crawl out of the stream. Oddly, alongside us we found dozens of *Abedus* doing exactly the same thing. Then I remembered: for reasons that nobody understood at the time, the giant water bugs were always the only critters remaining in streams after flash floods. At that moment it all clicked into place. Rainfall had to be the cue that warned them about an impending flash flood. No one had seen this flood escape behavior before. After all, who stands around in flash-flooding streams during monsoon rainstorms?

Later, we returned to the same canyon for experiments. By spraying pools with stream water from a fire hose, we imitated the rainstorm event and watched *Abedus* scurry to escape the stream again. We waited and discovered that *Abedus* return to the stream within a day after the floodwaters recede. Later, we collected *Abedus* from several desert streams in the area and compared escape behaviors among populations from different streams. Their responses were not the same. Individuals from streams where floods were common left the stream after about twenty minutes of rainfall, but individuals from streams where floods were rare needed much longer rainfall events to be convinced—some refused to leave the stream at all. It seemed that over hundreds or even thousands of years, evolution had fine-tuned these isolated populations so that their behaviors suited the flood regime of individual streams.

With clever strategies like these, perhaps it isn't surprising that *Abedus* adults are long-lived. No one knows exactly how long, but we have recaptured individuals in the wild that we had marked two years before. In some ways, *Abedus* are more like frogs or salamanders than other kinds of aquatic insects. Although they are usually found in the water, both juveniles and adults breathe air and can live on land for periods of time, as we saw during the flood. Adults produce multiple clutches of offspring, often with different mates. Like other giant water bugs, *Abedus* males provide exclusive parental care by brooding eggs until they hatch. Females glue eggs to the backs of males, and the males carry eggs until they hatch, up to a month later. Males brood the eggs by keeping them just at the surface of the water, where they are exposed to air but remain moist.

Unlike most aquatic insects, including most other giant water bugs, adult *A. herberti* cannot fly. They have wings underneath the leathery shells that cover their abdomens but lack the wing muscles to use them. No one knows why this is so, but perhaps evolution has favored the loss of wing muscles (which are physiologically expensive to produce). In remote desert streams, finding another permanent habitat can be risky. If you live in a nice stream, perhaps the best strategy is to stay put, although we were impressed with the speed at which these bugs crawled overland to hide in the woods.

An interesting side effect of their flightlessness is that *A. herberti* are good indicators of perennial stream habitats. If you find them living in a stream, chances are that the stream never completely dries up, even if it shrinks down to a few stagnant pools during drought years. Genetic studies suggest that some populations have been confined to specific streams since the Pleistocene days of woolly mammoths and saber-toothed tigers, so some individual populations might also harbor a unique genetic heritage.

They may be attentive parents and clever escape artists, but are *A. herberti* "ferocious" water bugs? By injecting potent venom with their piercing mouthparts, these insect goliaths subdue large prey like fish, salamanders, and frogs. Compared to other giant water bugs, *Abedus,* who measure a mere four centimeters long, are considered only medium-sized, but they can eat prey many times their own size, and they cannibalize each other when food is scarce. To a hapless tadpole, the syringe-like piercing rostrum and raptorial (clasping) front legs must appear ferocious. Even an intrepid scientist might flinch at the idea of being injected with enzymes that rapidly dissolve tissue into liquid. However, despite handling hundreds of them, I've been bitten only a handful of times by what are actually quite gentle bugs, and the pain is far less than a bee sting. The common name "toe biter" also seems unwarranted—I have yet to witness a bona fide case of toe biting. Most reports can be attributed to sharp rocks, thorns, or naucorids (a much smaller aquatic bug that delivers an excruciatingly painful bite).

But don't take my word for it. If you find yourself in the land of the ferocious water bug, tread carefully. Watch your toes. But also keep a sharp eye out to observe these fascinating creatures.

Abedus herberti

Life cycle: Iteroparous (i.e., produce multiple clutches of offspring), with a long adult stage (up to 2 years, possibly more).
Larvae: 5 larval instars.
Adults: Up to 40 mm in length. Males brood eggs on their backs for up to 1 month.
Feeding: Predatory (piercing mouthparts) on insects, fish, frogs, and each other.
Habitat Indicators: Because they are flightless, these desert insects are good indicators of perennial stream habitat.

22 Riding the Current for the Riverine Backswimmer

Michael Bogan

Our raft was gliding nearly silently over the roiling, chocolate-brown water, when I noticed a slight disruption on the left side of the boat. We were on the third day of an eight-day rafting trip down one hundred and twenty miles of wild river in a roadless area of central Sonora, Mexico. The majority of our ragged group of eighteen people in four rafts and two kayaks were biologists of one variety or another; our goal was to document as many species as possible during the voyage. My specialty is aquatic insects, and one of the bugs I was searching for was a strange type of backswimmer known as *Martarega mexicana*. When I noticed the hundreds of tiny splashes on the left side of the boat, I knew that I'd found my bug.

Unlike most backswimmers commonly found in ponds and lakes around the world, *Martarega* wouldn't be caught dead in still water. This genus of backswimmer is exclusively found in

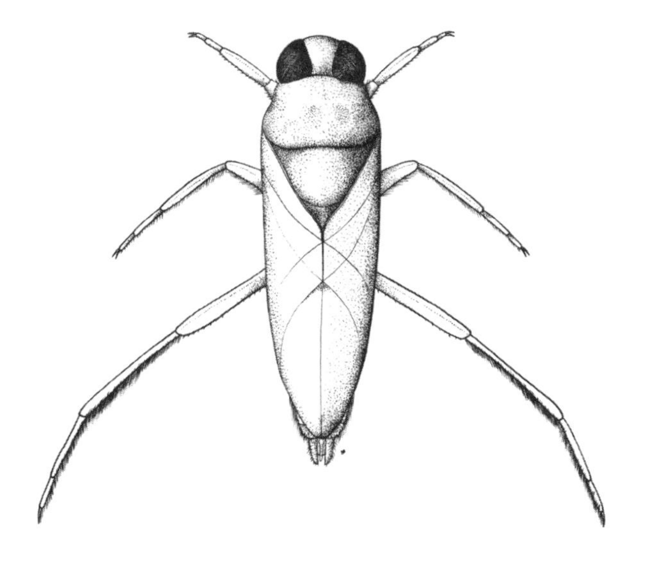

medium- to large-sized streams from the southwestern United States to central South America. In the runs and eddies of these streams, *Martarega* come together by the hundreds to form what have been described as "large, gregarious schools." They also have the unique habit, for aquatic insects anyway, of jumping out of the water when threatened by fish or avian predators—in our case they had been threatened by our large raft of biologists. As the current carried our raft downstream into the next set of whitewater rapids, I managed to swing my net through the leaping school of *Martarega* and toss some into a collection vial. Then I grabbed tightly to the boat as we plunged into the rapids.

The continuous downstream drive of the current was essential for our riverine voyage, and it is also essential in delivering meals to hungry beaks of *Martarega*. Unlike most backswimmers, *Martarega* do not have large, powerful rear legs capable of bursts of speed to capture mobile aquatic prey. Instead, *Martarega*'s hind legs have the same type of "low mechanical advantage" that horses do, perfect for long periods of sustained running (or swimming) instead of short sprints. This design allows *Martarega* to swim against a current without tiring, and in that current they wait for food to be carried to them. Sometimes they will eat aquatic insect larvae, but, when given a choice in an experimental study, their favorite food was ants! In most tropical and subtropical streams, there is likely no shortage of ants to accidentally fall out of the leafy stream canopy and get carried to the waiting arms of *Martarega*.

In addition to having leafy canopies, most subtropical streams also experience intense flow variation during the rainy season. It was this flow variation that allowed us to make our expedition down the river. Monsoon storms between June and September can deliver as much as three inches of rain per hour—this intense rain results in great increases in runoff to streams. The river we were rafting, the Rio Aros, is not much wider than a raft during the dry season, but during the monsoons it can swell to nearly the size of the Mississippi River. One morning we watched the river rise four vertical feet in less than an hour; we were therefore always sure to set up camp at least twelve vertical feet above the river's surface.

This intense flushing of upland and riparian habitats by the monsoon rains likely provides quite a buffet of terrestrial prey for *Martarega* in addition to the ants that fall from the canopy. While *Martarega*'s rear legs are relatively low-geared and made for long-distance swimming, their front legs have "high leverage coefficients," making them quite able to hold on to large, struggling insects. In one quiet reach of the Rio Aros, I watched through binoculars as a *Martarega* restrained and slowly killed a much larger grasshopper that made an ill-advised leap into the river. Like most aquatic hemipterans, *Martarega* inject a digestive enzyme into their captive prey and slowly digest them from the inside out. It was clearly not a pleasant end for the grasshopper.

A close examination of most *Martarega* individuals reveals another feature that distinguishes them from the rest of their backswimmer cousins: they are brachypterous, or short-winged. Most backswimmers in ponds and lakes have full-length (macropterous), powerful wings that are capable of long-distance flights. These fully functional wings serve them well in habitats like ponds, which may dry up seasonally, allowing them to fly in search of better habitats when local conditions get bad. In fact, when formerly dry ponds or stream pools fill up after a good rainstorm, backswimmers are among the first species to recolonize these habitats. However, *Martarega* live exclusively in perennial streams. Since these habitats don't normally dry up, *Martarega* can divert energy away from building large, powerful wings; instead, they use those resources for things like developing more eggs for reproduction.

So, why was I risking my life rafting on a wild Mexican river and scooping for *Martarega* before plummeting down Class IV whitewater rapids? Well, partly for a sense of adventure, partly to document the biological diversity of the area, and partly to solve a distributional puzzle that *Martarega* presented. Prior to the 1960s, *M. mexicana* was only known from a few streams between Mexico City and central Guatemala. Then, several biologists independently found *Martarega* in the Salt and Verde River drainages of central Arizona. These discoveries led to the question: how could there be a one-thousand-mile gap between populations of this species? Southern Arizona had been well surveyed for water bugs of all kinds, and *Martarega* had not been found. Surveys for aquatic insects in northern Mexico were much rarer. All collections had been made from highway road crossings of streams; however, no *Martarega* had been found in these surveys either. So, when I swiped my net through that disturbance on the left side of our raft and got a handful of *Martarega* (easily recognizable by their all-white color and eyes that touch on the top of their head), I was pretty excited.

Since finding *Martarega* on the rafting trip, I've managed to collect them from a number of places in northern Mexico, places that were seldom easy to reach but always spectacular. In Chihuahua, I was guided by Tarahumara Indians down some incredibly steep goat trails into what's known to Americans as Copper Canyon, an incredible labyrinthine system of volcanic canyons. Later, I backpacked with a brave friend for several days in the rugged gorge below Basaseachic Falls, a soaring nine-hundred-foot waterfall in Chihuahua, well deserving of its national park status. I drove with another brave friend on punishing dirt roads into the *narcotraficante* (drug-trafficker) country of northern Sinaloa, where locals living along the stream we surveyed made sure to tell us it was *not* safe to spend the night there. Each time, though, I put another "dot" on the map for *Martarega*, indicating that its distribution was not really that mysterious, rather that surveys in the appropriate locations had been lacking.

I have since found *Martarega* from Jalisco in central Mexico to within ten miles of the Arizona–Sonora border, but never between the border and the known populations in central Arizona. This remaining gap in the distribution of *Martarega* remains mysterious, but may result from deteriorated habitat conditions in southern Arizona. While perennial river habitats were, historically, uncommon in southern Arizona, several permanent streams should provide good habitat for *Martarega*, including the San Pedro and Santa Cruz rivers. However, declining water tables, stream diversions, and exotic invasive species during the last hundred years or so have been rough on most of these streams. Nearly all of the formerly perennial reaches have become ephemeral (that is, temporary) and unsuitable for the short-winged *Martarega*., The few remaining perennial streams, such as the San Pedro River, are now dominated by insect-devouring exotic invasive fish species, like green sunfish. All of the streams where I found *Martarega* in Mexico have unmodified stream flow regimes and few to no exotic fish species.

To me, the story of *M. mexicana* suggests a couple of things. First, we need more monitoring and survey efforts of streams in many areas of North America. *Martarega* is just one of any number of species that exhibit what are apparently "strange" distributions. As biologists and citizen stream monitors are able to survey more and more remote or previously overlooked streams, we will likely find that these distributions are not as disjunct or strange as they once seemed. Secondly, we need to better understand what drives the distributions of aquatic insect species. Most ecological and biomonitoring studies have focused on the most common EPT (mayfly, stonefly, and caddisfly) taxa, as these tend to be insects that are sensitive to human disturbance. However, the presence, absence, and/or abundance of many non-EPT taxa can also yield critical information about the health and history of a given stream. As for me, you can bet that I'll continue to use these two suggestions as excuses for going out to wild places and having more adventures in search of interesting bugs.

Martarega mexicana

Life Cycle: Unknown.

Nymphs: Small, up to 5 mm, look similar to adults. *M. mexicana*, like most hemipterans, goes through 5 nymphal instars.

Adults: Can live 1 year or more. Since most *M. mexicana* are short-winged (brachypterous), they are poor fliers and likely do not disperse far aerially.

Feeding: Nymphs and adults are predators. They have been observed eating aquatic insect larvae and terrestrial insects (e.g., ants) that fall to the surface of the stream.

Habitat Indicators: Lower and mid-elevation (200–2,000 m) perennial creeks and larger rivers appear to be the preferred habitats of *M. mexicana*. Exotic invasive fish species and habitat alteration may drive local extinctions of this species. The species is currently known from central Guatemala to north-central Arizona.

23 Secrets of an Infrequent Flyer

Mark P. Miller

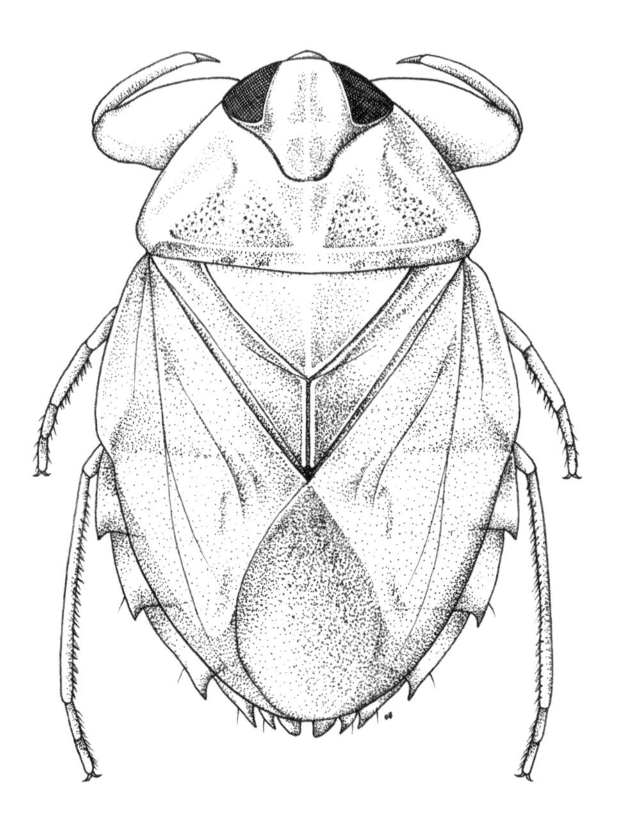

Streams of the Arizona White Mountains have marvelous aquatic insect assemblages. This high-elevation (5,000+ feet), heavily forested region in a remote section of eastern Arizona is an aquatic entomologist's dream. Textbook representatives of numerous well-known mayflies, caddisflies, stoneflies, and dragonflies are extremely abundant.

When I conducted fieldwork in the area, it seemed like every new rock we picked up held yet another new organism to collect. One of the true bugs I was studying, the one-centimeter-long *Ambrysus thermarum* (Hemiptera: Naucoridae) seemed most conspicuous, provided that we sampled correctly. After placing a kick net in a fast-flowing riffle and jostling some rocks around upstream, those stout, oval-shaped bugs were washed downstream by the current into our net. They were extremely active; as soon as we put our samples in a sorting tray, *A. thermarum* were swimming about wildly, seeking refuge from their exposed situation.

"Do they fly?" my companions asked on one of my first expeditions to the White Mountains.

"I don't know, but that's something I'd really like to try to find out," I replied. I'd already explained to my friends that I wanted to understand how aquatic insects disperse across the terrestrial environment. In essence, I wanted to develop a feel for how fragmented or connected different headwater streams are from the perspective of individual aquatic insect species. We already know, for many species, that juveniles disperse downstream with the current, and that adults disperse back upstream by flight (when they are capable). But do flying adults make extensive use of areas outside of the stream corridor? In particular, might *A. thermarum* fly to disperse? If so, then high rates of overland dispersal may lead to increased interactions among aquatic insect populations from different isolated headwaters.

My main study approach involved using genetic markers and population genetic analyses. If there were strong genetic differences within the stream network, then gene flow and dispersal were probably low. In contrast, weak genetic differences would indicate high gene flow and greater dispersal among populations. Conceptually, the approach is simple, and in practice, the approach is reasonably robust. However, using only markers and population analyses seemed slightly insufficient to me. I also wanted firsthand observations of adults flying through the air to help confirm patterns identified by genetic analyses.

I established a formal set of study sites in the mixed pine and spruce forests adjacent to the small streams that captured my interest. These second and third order streams were downstream of the headwaters (which we classify as first order streams). The stream sites were well shaded by adjacent forests. *A. thermarum* became one of the four species that I tried to find flying through the forest's open understory next to these streams. At the onset, I knew that this species could prove difficult to understand. Their morphology and life history provided no clear clues about how they dispersed. Juveniles are wingless and spend all of their time in the water. And like many other true bugs, adults have fully-formed wings. This attribute, at least superficially, suggested that the species is capable of flying over land.

However, naucorids display an unusual behavior that allows them to complete their life cycle entirely under water. While they require atmospheric oxygen to breathe, they have no need to leave the streams where they live to collect it. Instead, naucorids use an extremely specialized technique for storing air. Periodically, an adult will briefly extend the tip of its abdomen just above the water surface to gather a small air bubble that becomes trapped beneath its elytra, the rigid, modified forewings that cover and protect the hind wings. These air bubbles that naucorids collect are called sub-elytral air stores. Once replenished, the sub-elytral air store stays in contact with openings on the insect's abdomen (its spiracles) and allows the bug to breathe while submerged. Naucorids are efficient predators, and in the streams that I studied, there are ample soft-bodied prey items (other aquatic invertebrates) for them to consume throughout their life cycle. Food is not in short supply. Thus, adult *A. thermarum* hunt and feed exclusively within the stream; even copulation and egg laying occur completely underwater.

Given the lifestyle of these bugs, I addressed my initial question: do they fly? I used two different approaches—one scientific, and one decidedly unscientific—to gauge flight ability of my study species. My formal, rigorous assessment of flight involved setting up arrays of large "sticky traps" along transects that ran perpendicular to the stream channel. These were strips of cardboard painted with very gummy, and messy, glue. Friends, volunteers, and I carried these traps up and down the steep slopes of the forest alongside my study sites, attaching them to tree trunks at varying distances from the stream. The principle was simple: any insect that happened to land on one of these traps became stuck immediately. They became clear, indisputable evidence documenting the species's movement through the terrestrial environment. Over the course of this investigation, I caught numerous caddisfly, mayfly, stonefly, dragonfly, and aquatic dipteran (true fly) species in my traps. Some were detected in high numbers at reasonably large distances from the streams. But interestingly, no *Ambrysus* were ever detected, not even at distances less than five meters from the stream!

My "unscientific" flight assessment approach was simple. I collected a large number of *A. thermarum* adults from the stream, then, to induce them to fly, I cast them one by one up into the air. Most insects, when subjected to this type of harassment, will quickly orient themselves in the air and fly away. At the very least, they might use their wings to ease their fall. But not *A. thermarum*: when tossed into the air, they plummeted back to the ground like small stones.

These two lines of evidence initially led me to believe that *A. thermarum* simply did not fly; however, soon I was faced with a third, and conflicting, line of evidence. My genetic analyses revealed no genetic differences among the *A. thermarum* populations that I was studying. In fact, the genetic structure patterns closely resembled those from the caddisfly *Gumaga griseola*, a species that I frequently caught in my sticky traps; I knew from firsthand observation that *G. griseola* was a good disperser. This discrepancy was troubling, and it required a fair amount of time and thought to reconcile. Was I overlooking something? Was one of the lines of evidence somehow unreliable? Other researchers have reported seeing naucorids flying to lights at night or finding them in odd locations that could have only been reached through flight. Thus, they may actually disperse overland, but only at the right time and under the right circumstances. Though sometimes I found *A. thermarum* in unusual locations in the Arizona White Mountains, most of my study sites were permanent, fast-flowing streams that supported large *A. thermarum* populations. Those streams naturally had been the focus of my investigations and where I established my terrestrial sticky trap transects.

Luckily, occasional discoveries of *A. thermarum* in smaller intermittent streams that dried up in the summer told me more about its likely movement patterns. One of these uncommon observations was at a stream that appeared to be little more than a roadside ditch. Originally *A. thermarum* were quite abundant, but only at a section where the flow of water was relatively fast compared to the adjacent stream pools. As the season went on, the stream started to dry up, and my naucorids started to disappear too. It turns out that some aquatic Hemipterans can detect changes in stream flows or water levels; if either becomes too low, then it's a cue that their environment is changing and they'd best disperse to a new location. This was apparently the strategy used by *A. thermarum*.

Thus, I deduced that the network of smaller intermittent streams might actually be the prime locations for flying *Ambrysus* in the Arizona White Mountains. I had likely missed them in my traps because I had focused more on the picturesque, fast-flowing streams that initially captured my attention. Genetic analysis told me those bugs moved around, but I needed field studies to reveal their surprising dispersal strategy. And more importantly, to really understand their dispersal and flight tendencies, I needed to pay close attention to the extreme variation in habitats used by this infrequent flyer.

Ambrysus thermarum

Life cycle: Probably univoltine.

Nymphs: Oval-shaped; 5 instars with direct development to winged adult form (no pupal stage).

Adults: 8–10 mm long, and closely resembling juvenile forms (with the addition of developed wings and elytra).

Feeding: Juveniles and adults are predators of other macroinvertebrates. They capture and subdue their prey using their strong, sharp, raptorial forelegs.

Habitat indicators: Generally found in fast-flowing, medium-sized streams. Though the species' name may suggest a preference for warm waters, they are also readily found in colder, high-altitude streams.

Anatomy of a Mayfly

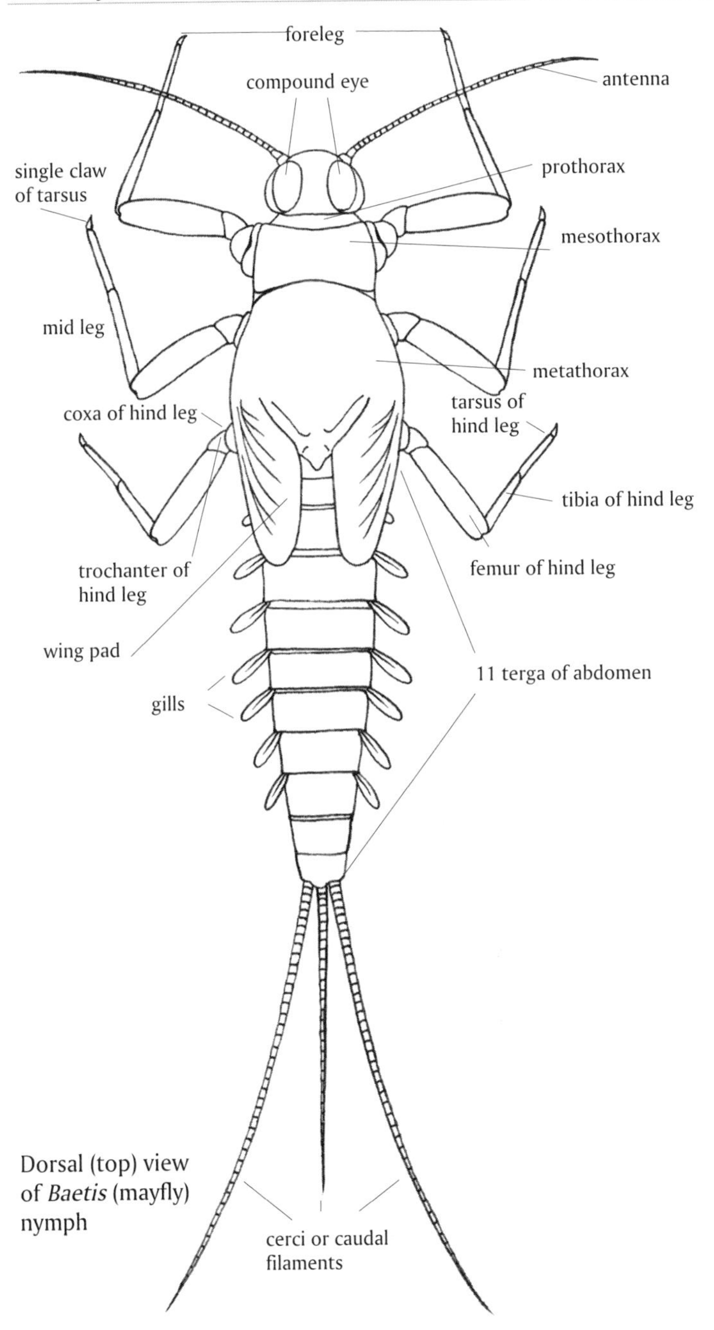

Dorsal (top) view of *Baetis* (mayfly) nymph

Biological Assessment: Using Biological Indicators to Evaluate the Health of a Waterbody

Why is biological assessment of a waterbody important?

Because aquatic organisms are adapted to the environments in which they live, their long-term exposure to what is happening in their individual habitats can give telltale signs of waterbody health. For instance, sensitive organisms will not withstand slugs of chemicals or other human activities that alter their habitats. These organisms are considered indicators because they would be eliminated or forced to move if they were exposed to adverse conditions. Aquatic organisms are analogous to the canaries that miners took into the mines with them many years ago; the canaries' negative reactions to lowered oxygen were warning signs to the miners. Similarly, we use biological indicators by comparing what we might expect in an ideal, healthy system to the absence or reduced numbers of aquatic organisms at sites we are studying; this information provides warning signs about the condition of a water resource.

In contrast to sensitive organisms, tolerant species (or taxa) adapt readily to changes in habitat or water quality; sometimes the abundance of these organisms can increase dramatically while more sensitive ones disappear. The concept of biological assessment, known as bioassessment, is based upon the presence or absence of expected taxa, proportions of sensitive or tolerant organisms, diversity, and abundance. Invertebrates, like those in the stories of this book, are excellent indicators of water quality for many reasons: they don't move around very much (relative to fish and larger organisms), they represent a broad range of sensitivity to the many things that may pollute or alter an aquatic environment, and they are easy to collect.

Biological assessment can be a proactive or reactive exercise. In the former, we might ask questions such as,

"What is the condition of my river?", or

"How do I know what should be in my stream before the planned housing development begins?"

Questions that are reactive would be,

"What is killing all of the fish in my stream?", or

"Why does my lake smell so bad?"

For all of these questions, collecting invertebrates in a bioassessment can be helpful. Identifying these organisms and recording abundance will help assess the condition of a waterbody. By contributing to a long-term base of information, these data will help measure changes (good or bad) over time.

How is a biological assessment designed?

There are many different designs for a bioassessment, ranging from simply sampling a stream in your backyard to a monitoring program of an entire river basin, lake system, or extending to the waters within a county, state, or region. Broad monitoring programs are usually conducted by water quality agencies. Their mandates are to assess the conditions of their jurisdictional waters, protect the designated quality level, identify where pollution or other perturbation is occurring, and determine what to do about it.

Other designs for bioassessment are more local and usually targeted to a specific location, area, or watershed. These designs usually incorporate a few sites that are spread out around a suspected source of pollution or an area of community interest such as a park, a neighborhood, etc. The various sites may be similar or show gradients of condition (for example, they can range from very good to pretty bad).

In both of these comparisons, knowledge of a waterbody's expected conditions is needed. This is determined by sampling "reference" sites—nearby locations that have not been affected by any major land development or sources of pollution. Once we know what occurs in these reference sites, we have a pretty good idea of what should be found in similar waterbodies within that region. In the United States, water quality agencies use bioassessments regularly and have developed a system of reference sites for waterbodies within their jurisdictions.

Communicating with appropriate water quality agencies can help you locate reference sites, and will help in various ways as you conduct bioassessments.

What kinds of methods are used for conducting bioassessments?

Methods for conducting bioassessments of invertebrates are typically very simple. They normally consist of using nets in streams and along the shores of rivers or lakes. Generally, a bioassessment aims to look at all of a system's invertebrates in order to capture the biodiversity, or the composition, of the community. Methods may focus on those habitats that are most likely to be inhabited by invertebrates, such as rocks and stones, wood and roots in water, and other vegetation.

Sampling usually involves standardized techniques for collecting material from the bottom, or benthos, of the water body; invertebrates can be sorted out of coarse debris and sediments using sieves at the collection site. Specialized sampling gear, such as grab samplers (dredges) can be used for deeper water. Sometimes, artificial substrates are placed in the water and left for several weeks; when these traps are retrieved, the invertebrates that colonized the substrates are collected. As you know from reading the stories in this book, some indicators have specific habitat requirements, so collection for these species should be sampled with appropriate habitats in mind. Once they are collected, invertebrates are best identified in the laboratory using a microscope or magnifier. Sometimes a more expert or trained individual will be able to identify the organisms in the field. However, better quality control is achieved by using a standard laboratory practice of processing, identifying, and cataloging the samples in a consistent manner.

When all of the organisms in your sample have been identified, then the chore of interpreting the kinds and numbers can begin. Analysis techniques that have been effective for assessing biological conditions summarize the data as "metrics." These are calculated terms or counts of organisms representing some measurable aspect of biological assemblage or function. Metrics characterize the biota and change in some predictable way with increased human influence. For instance, the number of taxa that are mayflies is a metric. The total number of taxa found in a sample is a metric. Other metrics represent the proportions of particular organisms, such as the percentage of stoneflies, or the percentage of shredders (those that feed upon leaf litter in the water). Metrics work as bioindicators because they are aggregates of taxa or organisms within a taxon that represent sensitivities or tolerances to pollution or other stressors. Though bioindicators can be individual taxa as discussed in these stories, they can also be a metric that integrates information from several taxa or organisms.

How do I learn more about bioassessment?

Most water quality agencies have information and guidance on their Web sites, or they have published protocols and other documents that describe their bioassessment design, methods, and results. Within the United States, the federal agencies of the U.S. Geological Society and the U.S. Environmental Protection Agency have well-established documents on bioassessment. State water quality agencies, as well as provincial governments in Canada, have also developed protocols and guidance. In North America, bioassessment procedures have been better developed for flowing waters than for lakes and wetlands; procedures have also been developed for estuarine and coastal areas. However, contacting the local aquatic biologist within your state, county, or provincial jurisdiction is likely the best place to start to benefit from information already in place.

Glossary

Anterior: toward the front, or head end, of an insect's body.
Asynchronous emergence: occurs when male and female adults of an insect species emerge over a long interval (e.g., weeks).
Benthos: organisms living on the bottom of aquatic systems. Also refers to those living on submerged surfaces, including vegetation.
Bioassessment: a comparative evaluation of an environment using biological measures with standardized metrics for appropriate biological groups (e.g., plants, invertebrates, and fish in streams) and associated physical and chemical conditions (e.g., temperature, nutrients, and turbidity in streams).
Biofilm: a complex, mucilaginous matrix made of algae, bacteria, fungi, and microscopic animals, which can coat submerged substrates (e.g., rocks, wood, or plants).
Bivoltine: refers to species that have two generations per year.
Boulder: stream rock about the size of a basketball or larger.
Carapace: an invertebrate's outer covering and external skeleton to which muscles are attached. It is shed periodically to allow for growth and development.
Cobble: stream rock about the size of a grapefruit.
Diatoms: a group of microscopic algae that have an external structure made of silica.
Diapause: a condition of arrested development or dormancy that occurs in a part of an insect's life history that is particular to the insect species. Can be induced by environmental or seasonal cues.
Dorsal: the top surface of an organism's body.
Elytra: hardened forewing of beetles and true bugs.
Ephemeral streams: streams that dry out annually, often in the summer.
Emergence: the time in an aquatic insect's life history when larval stage enters the terrestrial environment either to molt for a final stage to subadult (mayflies) or adult stage. Some insects emerge out of the water as adults.
Entomologist: a scientist who studies insects.
Exuvium *(plural, exuvia)*: an invertebrate's outer covering that is shed periodically and replaced by a new cuticle covering. Often found in stream drift or stream edges after insects molt in order to grow or emerge as adults.
Gravel: coin-sized small rocks in a stream larger than sand, but smaller than cobbles.
Hemimetabolous: a pattern of insect life history in which immature forms molt through multiple gradual changes during which the

juvenile resembles the adult. Wing pads appear in last stages of metamorphosis.

Holometabolous: a pattern of insect life history in which there are four distinct stages: egg, larval, pupa, and adult.

Imago: final winged adult stage. Mayflies have a unique pre-adult, winged form called a subimago.

Instar: stage of juvenile development that follows hatching from egg and continues until pupation, sub-adult, or adult stage, depending upon developmental pattern of the insect.

Intermittent stream: a stream that does not flow year round but instead dries up during part of the year.

Iteroparous: reproductive ability to produce multiple clutches of offspring.

Larva: immature, or juvenile, stage of insects. Also referred to as nymph or naiad.

Niche: environmental conditions within which an organism can survive and reproduce.

Nymph: the larval stage of hemimetabolous insects; often used in reference to dragonflies, damselflies, stoneflies, and mayflies.

Meniscus: concave- or convex-curved upper surface of a column of liquid, caused by surface tension.

Midges: common name for members of the Chironomidae family of true flies (order Diptera).

Multivoltine: refers to species that have two or more generations per year.

Perennial streams: streams that contain flowing surface water throughout the year.

Plankton: organisms that live in the water column, generally in non-flowing water bodies such as lakes or wetlands. While they exert some movement up and down, or sometimes between sand grains on the bottom of a water body, most plankton do not swim well.

Pupa: stage in holometabolous insects between larval and adult stage when major morphological changes occur.

Riparian: the ecosystem including a stream and the terrestrial environment adjacent to it. Influence of the riparian area can extend to a distance equaling the height of tallest riparian trees.

Rostrum: a beak-like structure that is part of the head of some insects. In true bugs (Heteroptera) the rostrum are mouthparts modified for sucking and piercing. In that group the rostrum is hidden in a groove when it is not feeding.

Semivoltine: refers to species that have one generation every two years.

Stream order: a term that signifies stream size. Headwaters are first order streams, and as subsequent streams of equal size join up, they are classified as the next size up (e.g., two first order streams make a second order stream, two seconds make a third, etc.).

Subimago: developmental stage of mayflies that occurs after nymphs emerge from the water prior to adult stage; also the time when reproductive organs are developing.

Synchronous emergence: occurring all at one time. Refers to emergence of aquatic adult insects occurring within a very short time.

Tarsus *(plural, tarsi)*: in aquatic insects, this is a segmented portion of the leg most distant from the body. It is divided into three to five small segments and has a clawed segment at the end.

Understory: vegetation, often brushy, growing under a forest canopy.

Univoltine: refers to species that have one generation per year.

Veliger: early life stage in some mollusks that live in the water column.

Ventral: the surface on the underside of an organism's body.

References for these definitions:

Hauer, R., and G. Lamberti. 2006. *Methods in Stream Ecology*, second edition. Amsterdam: Elsevier.

Resh, V. H., and D. M. Rosenberg. 1984. *The Ecology of Aquatic Insects.* New York: Praeger.

Resh, V. H., and R. T. Carde. 2003. *Encyclopedia of Insects.* Amsterdam: Academic Press.

Stylurus larva lying in wait for a crane fly larva. Drawing by Rob Cannings.

Useful References

Standard Keys for Identification

Edmunds, G. F., S. L. Jensen, and L. Berner. 1976. *The Mayflies of North and Central America.* Minneapolis, MN: U. of Minnesota Press.

Lehmkuhl, D. M. 1979. *How to Know the Aquatic Insects.* Dubuque, IA: W.M.C. Brown.

Merritt, R., K. Cummins, and M. Berg. 2008. *An Introduction to the Aquatic Insects of North America.* 4th edition. Dubuque IA: Kendall Hunt.

Stark, B. P., S. W. Szczytko, and C. R. Nelson. 1998. *American Stoneflies: A Photographic Guide to the Plecoptera.* Columbus, OH: Caddis Press.

Stewart, K. W., and B. P. Stark. 1993. *Nymphs of North American Stonefly Genera (Plecoptera).* Denton, TX: U. of North Texas Press.

Thorpe, J. H., and A. P. Covich. 2010. *Ecology and Classification of North American Freshwater Invertebrates.* New York, NY: Academic Press. 3rd edition.

Wiggins, G. B. 1996. *Larvae of the North American Caddisfly Genera (Trichoptera).* 2nd edition. Toronto: U. of Toronto Press.

Of General Interest

Allan, J. D. 1995. *Stream Ecology: Structure and Function of Running Waters.* London: Chapman and Hall.

Hauer, R., and G. Lamberti. 2006. *Methods in Stream Ecology.* 2nd edition. Amsterdam: Elsevier.

Resh, V., and D. Rosenberg. 1984. *The Ecology of Aquatic Insects.* New York: Praeger.

Useful Web Site Content

Freshwaters Illustrated (http://www.freshwatersillustrated.org) contains extensive photographic and extraordinary video resources for educators, scientists, and conservationists.

The Web site of the Society for Freshwater Science (formerly North American Benthological Society) (available at http://www.benthos.org) includes an online image library of aquatic invertebrates.

For details of specific aquatic invertebrates at risk, see information available at the non-profit Xerces Society for Invertebrate Conservation (available at http://www.xerces.org).

To Develop Methodologies of Bioassessment (with cursory keys and guides for developing bioassessment scores)

Adams, J., M. Vaughan, and S. Hoffman Black. *Stream Bugs as Biomonitors: A Guide to Pacific Northwest Macroinvertebrate Monitoring and Identification.* CD-ROM. Xerces Society (available at http://www.xerces.org).

Barbour, M. T., J. Gerritsen, B. D. Snyder, and J. B. Stribling. 1999. *Rapid Bioassessment Protocols for Use in Streams and Wadeable Rivers: Periphyton, Benthic Macroinvertebrates and Fish.* 2nd edition. EPA-841-B-99–002. U. S. Environmental Protection Agency, Office of Water.

Rosenberg, D. M., and V. H. Resh, eds. 1993. *Freshwater Biomonitoring and Benthic Macroinvertebrates.* New York: Chapman & Hall.

For the General Public

Hafele, R. 1981. *Complete Book of Western Hatches: An Angler's Entomology and Fly Pattern Field Guide.* Portland: Frank Amato Publications.

Hafele, R., and S. Roederer. 1995. *An Angler's Guide to Aquatic Insects and Their Imitations for All North America.* Boulder, CO: Johnson Books.

Hafele, R., and S. Hinton. 1996. Guide to Pacific Northwest Aquatic Invertebrates (Aquatic Biology Series). Portland; Oregon Trout. (Available from Oregon Department of Environmental Quality.)

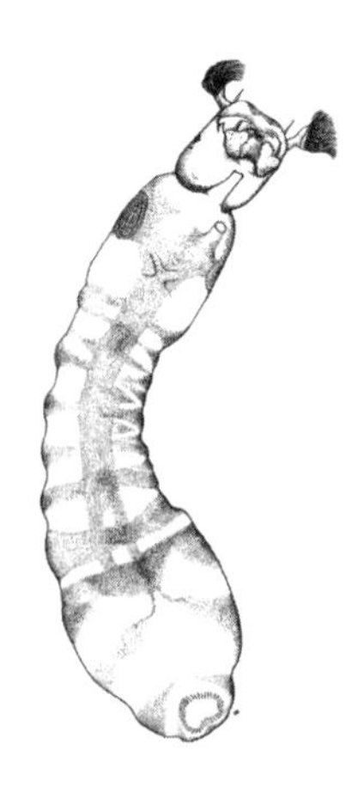

About the Contributors

Norman H. Anderson is professor emeritus of entomology at Oregon State University, where he taught a course in aquatic entomology for over thirty years. His research interests focus on the biodiversity and life histories of invertebrates, especially in small-stream habitats of the Pacific Northwest. In retirement Dr. Anderson has continued with this research by monitoring the fauna of summer-dry streams in his backyard for fifteen years.

Michael T. Barbour is director of the Center for Ecological Sciences at Tetra Tech, Inc., in Owings Mills, Maryland. Dr. Barbour serves as a technical expert to the U.S. Environmental Protection Agency for using biological indicators to assess the ecological condition of surface waters. He is particularly known for his publication of *Rapid Bioassessment Protocols for Streams*, which has served as a basis for implementing bioassessment in water quality agencies. He has also published two environmental mystery novels, *The Kenai Catastrophe* and *Blue Water, Blue Island*, as well as two children's books, *Caitlyn and Craig the Crayfish* and *Douglas Visits the Seashore.*

Christopher Beatty is a lecturer in ecology and evolution at Santa Clara University, California. Dr. Beatty studies life history evolution, behavior, and speciation, using dragonflies and damselflies as model organisms. In recent work he has studied territoriality, mimicry, and mating ecology in damselflies in Fiji, Peru, and the Azores Islands. In 2007 he received a PhD from Carleton University, Ontario, and was a postdoctoral fellow at the University of Vigo in Spain. His international studies resulted in book chapters on dragonflies and climate change, and island parthenogenesis and damselflies.

Fred Benfield is professor of ecology in the department of biological sciences at Virginia Tech in Blacksburg, Virginia, where he has been conducting research in southern

Appalachian streams for over forty years. His research focuses on responses of streams to historical and contemporary land disturbance, with a special interest in ecosystem-level processes and biodiversity. Dr. Benfield is a member of the Coweeta Hydrologic Laboratory Long-Term Ecological Research team that is investigating "ex-urbanization" (influx of population from the Charlotte–Atlanta corridor and beyond) of the southern Appalachian ecosystem in western North Carolina.

Michael Bogan is a staff research associate at the University of California, Santa Barbara's Sierra Nevada Aquatic Research Lab in Mammoth Lakes, California; he is also a PhD candidate in the department of zoology at Oregon State University. From 2000 to 2007, he worked for the California Department of Fish and Game in Bishop, California. Bogan has worked with progressively smaller aquatic organisms through the years, from trout to frogs to pupfish, finally settling on aquatic macroinvertebrates during his master's program at Oregon State. His research focuses on understanding how diverse communities and robust populations of aquatic insects persist in arid-land streams, despite severe flooding and droughts. He especially enjoys conducting aquatic biodiversity surveys in Mexico with his diverse team of American and Mexican collaborators.

Boonsatien Boonsoong is a lecturer in the department of zoology, faculty of science at Kasetsart University, Bangkok, Thailand. Dr. Boonsoong's major research interests focus on the rapid bioassessment of Thai streams with benthic macroinvertebrates and the taxonomy of mayflies, especially the family Heptageniidae. He teaches undergraduate courses on general zoology, invertebrate zoology, ecology, pollution biology, and biological drawing. He is co-author and illustrator of *Identification of Freshwater Invertebrates of the Mekong River and Tributaries*, and he has developed a citizen biomonitoring procedure to help educate the people of Thailand about environmental awareness.

Rob Cannings has been the curator of entomology at the Royal British Columbia Museum since 1980; from 1987 to 1997 he led the museum's natural history section. His research interests focus on insect systematics and faunistics, especially in the Odonata (damselflies and dragonflies) and Asilidae (robber flies), but he publishes widely on many insect groups. Dr. Cannings is the author or co-author of five books, including *Birds of the Okanagan Valley, British Columbia* (1987), *The Dragonflies of British Columbia* (1977), *Introducing the Dragonflies of British Columbia and the Yukon* (2002), and *The Systematics of Lasiopogon (Diptera: Asilidae)* (2002), and he has written many scientific and popular articles, mostly on insects and birds. He is active on the scientific committee of the Biological Survey of Canada and is a member of the arthropod subcommittee of the committee on the status of endangered wildlife in Canada.

Lynda D. Corkum is a professor of biological sciences at the University of Windsor and past president of the International Association for Great Lakes Research. Her research interests include the ecology and behavior of fishes and aquatic insects. Dr. Corkum has written a field guide, *Fishes of Essex County and Surrounding Waters*, which includes a detailed account of fishes in Lake St. Clair, the Detroit River, and Lake Erie.

Gregory W. Courtney is professor of entomology at Iowa State University in Ames. His major research and teaching interests are insect systematics and aquatic entomology. Specific areas of study include the morphology, systematics, biogeography, and ecology of aquatic Diptera, the phylogeny of dipteran families, and the biodiversity and conservation of aquatic habitats. Recent projects have focused on the fauna of torrential streams in Asia (especially Nepal and Thailand), Australasia (Australia and New Zealand), and the southern Andes (especially Patagonian Argentina and Chile). Dr. Courtney also is curator of the Iowa State Insect Collection and director of Iowa State University's Insect Zoo.

Deb Finn is presently a postdoctoral fellow at the University of Birmingham (U.K.). After receiving her PhD from Colorado State University in 2006, she has studied stream ecology around the world through postdoctoral positions not only at Birmingham but also at Oregon State University, the Swiss Federal Institute of Aquatic Sciences and Technology, and the Australian Rivers Institute at Griffith University. Dr. Finn is interested in spatial distributions of stream insects and their habitat, how spatial patterns of stream biodiversity respond to climate change, and how science can inform freshwater conservation. To address these topics, she uses a variety of methods, including genetics, geographical information systems, and extensive sampling in a wide variety of stream types, from hot deserts to the alpine tundra. She enjoys teaching aquatic ecology to students and non-scientists and writes the newsletter for the Society for Freshwater Science.

Donna Giberson is a professor of biology at the University of Prince Edward Island in Charlottetown, Canada, where she has taught since 1992. Her main research interests are in aquatic insect faunistics and in the taxonomy of mayflies and stoneflies, for which she has current projects in streams of eastern Canada and the Canadian Arctic. Dr. Giberson is also working with school and community groups in the Northwest Territories and Nunavut to help involve local youth in using aquatic insects in water quality assessment and biodiversity studies. Dr. Giberson has a keen interest in scientific literacy and is co-manager of a program in the UPEI biology department aimed at improving the writing and reading level of university undergraduates.

Judith Li is an associate professor (retired) in the department of fisheries and wildlife at Oregon State University. Her primary research interests focus on riparian food webs, particularly the role of invertebrates (aquatic and terrestrial) in the diet of fish and riparian birds in arid as well as wetter ecosystems of the Pacific Northwest. Many of her studies focus on the effects of forestry and agricultural activities on small streams. Dr. Li has received several university and national teaching awards, particularly for her interests in multicultural diversity. Besides her research activities, she continues her interests in educating the public and K-12 teachers. In 2007 she edited a volume on cultural ecology entitled *To Harvest, To Hunt; Stories of Resource Use in the American West*, and she currently has in press a children's science book, *Ellie's Log*.

David A. Lytle is an associate professor in the department of zoology at Oregon State University. He is broadly interested in how organisms evolve strategies for surviving extreme events, such as floods and droughts. Dr. Lytle has studied this in habitats ranging from desert tinajas the size of kiddie pools to large dammed rivers with experimental flood releases. In addition to conducting research projects, he teaches courses in aquatic entomology and evolutionary biology.

Richard W. Merritt is professor of entomology at Michigan State University in East Lansing. Dr. Merritt's major research interests focus on aquatic insect feeding ecology, animal–microbial interactions, population dynamics, and influence of environmental factors on immature larvae, especially the Diptera. His most recent research has concentrated on the ecology of Buruli ulcer, a neglected emerging disease in Africa that involves aquatic insects as potential reservoirs of disease, in addition to researching the biomonitoring of streams and rivers and the effects of pollutants on aquatic ecosystems. He is co-editor of *An Introduction to the Aquatic Insects of North America*.

Mark P. Miller is a statistician with the U.S. Geological Survey (Forest and Rangeland Ecosystem Science Center) in Corvallis, Oregon. Though his "official" job title would not seem to reflect it, Dr. Miller is a broadly trained ecologist and population geneticist. His work largely focuses on use of molecular genetic techniques to address diverse topics in wildlife conservation and management. In addition to studying aquatic insects, he has investigated a wide variety of taxonomic groups including birds, fish, mammals (including humans), mollusks, amphibians, and fungi. He also develops and distributes statistical software for the analysis of population genetic data (http://www.marksgeneticsoftware.net).

Marilyn Myers is a biologist with the U.S. Fish and Wildlife Service in Anchorage, Alaska. Her doctoral research investigated aquatic invertebrates in island habitats, studying not only traditional oceanic island habitats, but also considering desert springs, islands in arid landscapes. As an endangered-species biologist, she evaluates species ranging from spring snails to marine mammals for inclusion on the threatened and endangered species list and works to ensure that listed species receive the protections afforded to them under the Endangered Species Act.

Dave Penrose retired from North Carolina State University and the North Carolina Division of Water Quality in 2008 and moved to the beautiful mountains of western North Carolina. Before his retirement, much of his work was in assessing the ecological functions of restored stream features and attempting to develop a protocol for assessing the regulatory effectiveness of restoration projects. He also became very active in research to determine stream origins in North Carolina. However, retirement has proven difficult for Dr. Penrose, who continues to work on restoration projects throughout the southeast, teaches workshops on the taxonomy of aquatic insects, and is active on both state and national policy review teams.

Vincent Resh is professor of entomology in the Department of Environmental Science Policy and Management at the University of California, Berkeley. His research for the past forty years has been on biological monitoring of streams and rivers, life history studies of aquatic insects, and general aquatic ecology. Dr. Resh has been an advisor to The River Blindness (*Onchocerciasis*) Control Program in West Africa for fifteen years and to the Mekong River Commission for eight years. He is the co-editor of books on aquatic insect ecology and biomonitoring, including *The Encyclopedia of Insects*.

John Richardson is a professor at the University of British Columbia, Canada, in the Department of Forest Sciences. Dr. Richardson has over thirty years of experience studying small streams and their animals, including invertebrates, amphibians, and fish. His research team studies the human impact of forestry and other land uses on streams, while examining the roles and population dynamics of particular species. He has also conducted extensive studies on the flow of energy and materials across ecosystem boundaries, particularly the inputs of leaf litter to streams, and considered how these linkages influence aquatic ecosystems. Dr. Richardson teaches courses in freshwater biology and wildlife ecology. He is a member of several endangered species recovery teams and is director of the Stream and Riparian Research Laboratory.

Michael C. Swift is visiting assistant professor of biology and environmental studies at St. Olaf College. He has studied zooplankton ecology and photophysiology, aquatic toxicology, and stream ecology throughout the U.S. and western Canada. He teaches introductory biology, invertebrate biology, ecology, limnology, environmental studies, and special topics courses for non-biology majors. Periodically Dr. Swift also teaches aquatic ecology at the Coe College Wilderness Field Station beside the Boundary Waters Canoe Area Wilderness in northern Minnesota. He co-supervises the St. Olaf Biology in South India Program and was a Fulbright-Nehru Visiting Lecturer in Tamil Nadu, India from 2009 to 2010. He maintains a passionate interest in the biology of the phantom midge, *Chaoborus.*

Mark Vinson is a researcher with the U. S. Geological Survey, Great Lakes Science Center, Lake Superior Biological Station, in Ashland, Wisconsin. He has pursued aquatic invertebrates from Africa to the arctic. For many years he was director of the "Bug Lab" at the U.S. Forest Service Laboratory, Utah State University. He is currently studying the biota of Lake Superior and looking forward to future expeditions to faraway places.

John R. Wallace is professor of biology at Millersville University in Millersville, Pennsylvania. Dr. Wallace's major research interests focus on the population dynamics and feeding ecology of mosquitoes and other biting insects, invasive species ecology, the responses of macroinvertebrate communities to anthropogenic disturbances such as stream restoration, and riparian rehabilitation. He also investigates the role of aquatic organisms such as insects, algae, and crayfish in decomposition processes within forensic science. He is a fellow in the American Association of Forensic Sciences and is a board-certified forensic entomologist. He is co-editor of *Wildlife Forensics: Methods and Applications.*

David Wartinbee has been a professor of biology at Kenai Penninsula College for thirty-four years. His primary research interest is in the Dipteran family Chironomidae. Recently he identified eighty-eight species of midges found in the Kenai River. His teaching experience has been mainly in the medical arena, teaching human anatomy and physiology, but he periodically teaches graduate courses in stream ecology. He is also admitted to the bar for the practice of law in Pennsylvania and Alaska. Most recently he has been writing a science column for the *Redoubt Reporter*, mainly discussing streams and the organisms found around them.

John Woodling received a bachelor's of science from Southern Colorado State College in 1968 and a master's of science from the University of Louisville in 1971. He worked for the Colorado Water Quality Control Division as a research biologist from 1973 to 1978 and, from 1978 to 2003, has held several positions in the Colorado Division of Wildlife, performing field studies on streams and rivers in Colorado and in adjoining states. After earning a PhD from the University of Colorado in 1993, Dr. Woodling also taught stream biology at the University of Colorado's Boulder campus until retiring from the Division of Wildlife in 2003. Since 2003, he has moved to Grand Junction, Colorado, and spends time fishing, rafting, consulting for organizations involved in environmental protection, and doing artwork.

Index

Note: References in this index include exact phrases within the text and also references which describe a characteristic or process more broadly.

B

behavior
  alarm response, 121-22
  drift, 3, 16, 17, 21, 27
  escape reaction 124,136
  vertical migration, 44, 98
  grouping, 122, 128
  See also: crawlers; swimmers; clingers; burrowers; case building; flying; mating; prey detection; shredders
burrowers/burrowing, 3, 14, 17, 18, 21, 23, 50, 59, 101, 104, 107

C

case building, 34, 39-40, 54, 59, 62-63
clingers, 17, 18, 21, 72, 75
collectors
  filterers, 10, 36, 42, 47, 65, 71, 72, 73, 75, 78, 95
  gatherers, 14, 21
crawlers/sprawlers, 16, 17, 101

D

detritivores, 64
development
  complete, 35, 66, 113
  gradual, 2, 19, 23, 101, 115
diapause, 2, 23, 28, 35, 40, 59, 63, 88, 101
dispersal, 133
  beetles, 113, 120
  bugs, 115, 124, 129, 131, 133-36
  caddisflies, 35, 135
  dragonflies, 110
  stoneflies, 28, 31

dispersal (continued)
true flies, 73
See also flying

E

ecosystems
desert, 48, 123, 126
headwaters, 25, 60
lake/pond outlets, 44, 47, 73, 75, 77
lakes, 22, 91, 93, 95, 98, 111, 116, 119, 122, 127, 129
mountains, 25, 28, 51, 78, 85, 132
rivers, 10, 13, 14, 21, 30, 73, 82, 88, 104-107, 110, 111, 130, 131
springs, 50, 73
temporary streams, 48-50, 57, 60, 136
waterfalls, 81
egg laying
beetles, 122
bugs, 115. 125, 126, 134
caddisflies, 38-39, 45, 55, 58, 60, 62, 64
dragonflies, 106, 110
mayflies, 3, 7, 19
stoneflies, 23
true flies, 66, 77, 81, 87, 93, 98, 99
emergence
beetles, 113
caddisflies, 35, 40, 41, 45, 47, 54-55, 58, 60, 64
dragonflies, 101, 106, 107, 110, 111
mayflies, 1-3, 5-6, 8, 10, 13, 14, 18-19
stoneflies, 23, 29, 33
true flies, 66, 73, 78, 82, 86, 88, 91-92, 98
See also exuvia
exuvia, 6, 91-92, 104-6

F

flying/flight, 22-23, 28, 66, 87, 101, 113, 129, 133, 135, 136
flightless, 93, 125-26
flow. *See* habitat requirements, flow

functional feeding groups. *See* collectors, filterers; collectors, gatherers; grazers; predators, insects; shredders

G

generation time. *See* life histories
gills
  abdominal, 1, 14, 17, 18, 21, 34, 35, 100
  caudal, 100
  thoracic, 22
grazers (scrapers), 41, 56, 80, 82, 88, 112

H

habitat requirements
  cool water 22, 25, 30, 33, 34, 41, 60, 78, 88, 93
  flow
    fast/rapid, 17, 21, 25, 42, 45, 72, 82, 87, 88, 110, 111, 113, 133
    calm/slow, 17, 21, 42, 47, 56, 110, 113, 122
  perennial water, 125-26, 129- 31
  riparian vegetation, 23, 41, 45, 47, 55, 58, 100-102, 107, 122
  stable banks, 107
  well-oxygenated water, 21, 22, 33, 73, 82, 88, 93, 113
  See also sensitive taxa; tolerant taxa
halteres, 66

I

imago, 2, 6, 14, 19

L

life histories
  bivoltine, 18, 19, 35, 60, 66, 88, 91
  more than one generation/year (merivoltine, semivoltine), 23, 28, 31, 35, 92, 101, 107, 125
  multivoltine, 2, 66, 72, 88, 91, 99
  univoltine, 2, 10, 14, 19, 23, 35, 41, 46, 51, 55, 60, 64, 66, 72, 78, 81, 88, 99, 111, 122, 131, 136

M

mating
- bugs, 115, 125
- caddisflies, 45, 41, 47, 55, 58, 64
- dragonflies, 102, 106
- mayflies, 1-2, 5, 7, 10, 19
- stoneflies, 23
- true flies, 77, 78, 87, 93, 99

metamorphosis. *See* development

microhabitats
- bedrock, 54, 56, 78
- cobble/boulders, 9, 33, 56, 75-76, 78, 82, 85, 87, 110
- edgewaters, 21, 45, 46-47, 53, 58, 62, 64, 77, 93, 99, 100, 113, 116, 122
- gravels, 21, 23, 53, 60, 82
- moss, 17, 42, 55, 57
- mud, 14, 17, 21, 57, 93, 95, 99, 110-11, 113
- riffles, 7, 8, 21, 82, 86, 133
- sand, 10, 42, 50-51, 59-60, 92, 104-5, 107, 110-11, 113
- silt, 9-10, 14, 104-5
- vegetation, 23, 41, 45- 47, 55, 58, 71-73, 100-102, 122
- wood, 16, 21, 35, 45, 49, 58, 60, 63-64, 110, 122

molting, 1-3, 7, 10, 14, 19, 22-23, 33, 35, 40, 46, 54, 66, 101, 107

P

predation
- by birds, 29, 33, 52, 63, 91, 106, 128
- by fish, 5-8, 10, 14, 25, 29, 33, 34, 38, 54-55, 63, 91, 98-99, 106, 119-20, 128-29, 131
- by insects, 43, 70, 82, 93, 95, 98-99, 100-101, 107, 110-11, 112, 114, 122, 123, 126, 128, 131, 134, 136

prey detection by insects, 44, 96, 101-2, 120, 122

prolegs, 35, 65, 72, 84, 88

pupae/pupation
- beetles, 122
- caddisflies, 40, 46, 50, 54, 58, 60, 63
- true flies, 66, 73, 77, 81, 85, 87-88, 91-92, 98

R

respiration
beetles, 113
bugs, 115
caddisflies, 35, 59

S

sampling/samplers
aerial net, 5, 48, 58, 72, 108-9, 128
blacklight traps, 48
D-frame/kick nets, 49, 85, 133
dredge, 140
emergence trap, 48
hand picking, 16, 48
pipe, 49
soil sieve, 50, 140
spider webs, 87
sticky trap, 48, 134-35
Surber sampler, 48
zooplankton net/tow, 44, 94, 96-97
sensitive taxa, 138, 141
to chemicals, 3, 14
to sediments, 41, 47, 88
to pollution, 14, 21, 23, 33, 51, 66
to warm temperatures, 51, 78, 88
See also habitat requirements
shredders/shredding, 21-22, 25-28, 31, 58, 60, 62, 64, 141
silk, 34, 62, 76, 92
in case building, 34, 37, 39-40, 54, 58, 63-64
in pads, 72, 75
in pupal cocoons, 34, 46, 57-58, 64, 73
in tubes, 92
use by net spinners, 34, 42-44, 46-47
sprawlers. *See* crawlers
subimago, 2, 5, 6, 10, 14, 19, 35
swimmers/swimming, 17-18, 112, 119-20, 128

T

tolerant taxa, 138, 141
  to acid, 98-99
  to disturbance, 106-7
  to drought, 23, 35, 59-60
  to high nutrients, 36, 73
  to low oxygen, 93, 116
  to low water quality, 47
  to pollution, 66, 73, 93,
  to salinity, 90, 93, 106
  to warm temperatures, 47, 59, 107

V

voltinism. *See* life histories

Z

aooplankton, 44, 95-97